Materials and the Environment: Eco-Informed Material Choice

Materials and the Environment
Eco-Informed Material Choice

Michael F. Ashby

AMSTERDAM • BOSTON • HEIDELBERG • LONDON
NEW YORK • OXFORD • PARIS • SAN DIEGO
SAN FRANCISCO • SINGAPORE • SYDNEY • TOKYO

Butterworth-Heinemann is an imprint of Elsevier

Butterworth-Heinemann is an imprint of Elsevier
30 Corporate Drive, Suite 400, Burlington, MA 01803, USA
Linacre House, Jordan Hill, Oxford OX2 8DP, UK

∞ Recognizing the importance of preserving what has been written, Elsevier prints
its books on acid-free paper whenever possible.

Library of Congress Cataloging-in-Publication Data
Application submitted

British Library Cataloguing-in-Publication Data
A catalogue record for this book is available from the British Library.

ISBN: 978-1-85617-608-8

For information on all Butterworth-Heinemann publications
visit our Web site at www.books.elsevier.com

Printed in Canada

08 09 10 11 12 13   10 9 8 7 6 5 4 3 2 1

Table of contents

Preface

The environment is a system. Human society, too, is a system. The systems coexist and interact, weakly in some ways, strongly in others. When two already complex systems interact, the consequences are hard to predict. One consequence has been the damaging impact of industrial society on the environment and the ecosystem in which we live and on which we depend. Some impacts have been evident for more than a century, prompting remedial action that, in many cases, has been successful. Others are emerging only now; among them, one of the most unexpected is changes in global climate that, if allowed to continue, could become very damaging. These and many other ecoconcerns derive from the ways in which we use energy and materials. If we are going to do anything about it the first step is to understand the origins, the scale, the consequences, and the extent to which, by careful material choice, we can do something about it. And that requires *facts*.

The book

This text is a response. It aims to cut through some of the oversimplification and misinformation that is all too obvious in much discussion about the environment, explaining the ways in which we depend on and use materials and the consequences of their use. It introduces methods for thinking about and designing with materials when one of the objectives is to minimize *environmental impact*—an objective that is often in conflict with others, particularly that of minimizing *cost*. It does not aim to provide ultimate solutions—that is a task for future scientists, engineers, designers, and politicians. Rather, it is an attempt to provide perspective, background, methods, and data—a toolbox, so to speak—to introduce students to one of the central issues of environmental concerns, that surrounding the use of materials, and to equip them to make their own judgments.

The text is written primarily for students of Engineering and Materials Science in any one of the four years of a typical undergraduate program. It is organized in two parts. The first, Chapters 1 to 11, develops the background and tools required for the materials scientist or engineer to analyze and respond to environmental imperatives. The second, Chapter 12, is a collection of profiles of materials presenting the data needed for analysis. The two together allow case studies to be developed and provide resources on which students can draw to tackle the exercises at the end of each chapter (for which a solution manual is available) and to explore material-related eco-issues of their own finding.

To understand where we now are, it helps to look back over how we got here. Chapter 1 gives a history of our increasing dependence on materials and energy. Most materials are drawn from nonrenewable resources inherited from the formation of the planet or from geological and biological eras in its history. Like any inheritance, we have a responsibility to pass these resources on to further generations in a state that enables them to meet their aspirations as we now do ours. The volume of these resources is enormous, but so too is the rate at which we are using them. A proper perspective here needs both explanation and modeling. That is what Chapter 2 does.

Products, like plants and animals, have a life cycle, one with a number of phases, starting with the extraction and synthesis of raw materials ("birth"), continuing with their manufacture into products, which are then transported and used ("maturity"), and at the end of life, sent to a landfill or to a recycling facility ("death"). Almost always, one phase of life consumes more resources and generates more emissions than all the others put together. The first job is to pin down which phase involves the most consumption. Life-cycle assessment (LCA) seeks to do this, but there are problems: as currently practiced, life-cycle assessment is expensive, slow, and delivers outputs that are unhelpful for engineering design. One way to overcome these issues is to focus on the main culprits: one resource, *energy*, and one emission, *carbon dioxide, CO_2*. Materials have an *embodied energy* (the energy it takes to create them) and a *carbon footprint* (the CO_2 that creating them releases). So, too, do the other phases of life, and materials play a central role in these also. *Heating and cooling* and *transportation*, for instance, are among the most energy-gobbling and carbon-belching activities of an industrial society; the right choice of materials can minimize their appetite for both. This line of thinking is developed in Chapters 3 and 4, from which a strategy emerges that forms the structure of the rest of this book.

Governments respond to environmental concerns in a number of ways applied through a combination of "sticks and carrots," or, as they would put it, *command and control* methods and methods exploiting *market instruments*. The result is a steadily growing mountain of legislation and regulation. It is reviewed in Chapter 5.

As engineers and scientists, our first responsibility is to use our particular skills to guide design decisions that minimize or eliminate adverse eco-impacts. Properly informed materials selection is a central aspect of this task, and that needs data for the material attributes that bear most directly on environmental questions. Some, like *embodied energy* and *carbon footprint*, *recycle fraction* and *toxicity*, have obvious ecoconnections. But more often it is not these but mechanical, thermal, and electrical properties that have the greatest role in design to minimize eco-impact. The data sheets of Chapter 12 provide all of these properties. Data can be deadly dull. It can be brought to life (a little) by good visual presentations. Chapter 6 introduces the material attributes that are central for the material that follows and displays them in ways that give a visual overview.

Now to *design*. Designers have much on their minds; they can't wait for (or afford) a full LCA to decide between alternative concepts and ways of implementing them. What they need is an *eco-audit*—a fast assessment of product life phase by phase and the ability to conduct rapid "What if?" studies to compare alternatives. Chapter 7 introduces audit methods with a range of examples and exercises in carrying them out using the data sheets in Chapter 12.

The audit points to the phase of life of most concern. What can be done about it? In particular, what material-related decisions can be made to minimize its eco-impact? Material selection methods are the subject of Chapter 8. They form a central part of the strategy that emerged from Chapter 3. It is important to see them in action. Chapter 9 presents case studies of progressive depth to illustrate ways of using the materials. The exercises suggest more.

Up to this point the book builds on established, well-tried methods of analysis and response, ones that form part of, or are easily accessible to, anyone with a background in engineering science. They provide essential background for an engineering-based approach to address environmental concerns, and they provide an essential underpinning for studies of broader issues. Among these are questions of *sustainability* (perhaps the most misused word in the English language today) and *future options*, an attempt to foresee future problems and potential solutions. They are the subjects of the last two chapters of Part 1 of the book.

The final chapter is straightforward. It is an assembly of 47 two-page data sheets for engineering metals, polymers, ceramics, composites, and natural materials. Each has a description and an image, a table of mechanical, thermal, and electrical properties, and a table of properties related to environmental issues. These data sheets provide a resource that is drawn on in the text of the book, enables its exercises, and allows you to apply the methods of the book elsewhere.

The approach is developed to a higher level in two further textbooks, the first relating to mechanical design,[1] the second to industrial design.[2]

The CES software[3]

The audit and selection tools developed in the text are implemented in the CES Edu 09 software, a powerful materials information system that is widely used for both teaching and design. The book is self-contained; access to the software is not a prerequisite. The software is a useful adjunct to the text, enhancing the learning experience and providing access to data for a much wider range of materials. It allows realistic selection studies that properly combine multiple constraints and the construction of tradeoff plots in the same format as those of the text.

[1]Ashby, M. F., Materials selection in mechanical design, 3rd ed., Chapter 4, Butterworth Heinemann, 2005, ISBN 0-7506-6168-2. (A more advanced text that develops the ideas presented here in greater depth.)

[2]Ashby, M. F., and K. Johnson, Materials and design: the art and science of material selection in product design, Butterworth Heinemann, 2002, ISBN 0-7506-5554-2. (Materials and processes from and aesthetic point of view, emphasizing product design.)

[3]Granta Design, www.grantadesign.com.

Acknowledgments

No book of this sort is possible without advice, constructive criticism, and ideas from others. Numerous colleagues have been generous with their time and thoughts. I would particularly like to recognize the suggestions and stimulus, directly or indirectly, made by Dr. Julian Allwood, Prof. David Cebon, Dr. Patrick Coulter, Dr. Jon Cullen, Prof. David MacKay, and Dr. Hugh Shercliff, all of Cambridge University, and Prof. John Abelson and the students of the University of Illinois Materials Science and Engineering classes 201 (*Phases and Microstructures*) and 498 (*Materials for Sustainability*) (2008), who trialled and proofread the manuscript. Equally valuable has been the contribution of the team at Granta Design, Cambridge, responsible for the development of the CES software that has been used to make many of the charts that are a feature of this book.

Introduction: material dependence

1.1 Introduction and synopsis

This book is about *materials*: the environmental aspects of their production, their use, their disposal at end of life, and ways to choose and design with them to minimize adverse influence. Environmental harm caused by industrialization is not new. The manufacturing midlands of 18th-century

Renewable and nonrenewable construction. Above: Indian village reconstruction. (Image courtesy of Kevin Hampton, www.wm.edu/niahd/journals.) Below: Tokyo at night. (Image courtesy of www.photoeverywhere.co.uk index.)

England acquired the nickname the "Black Country" with good reason; to evoke the atmosphere of 19th-century London, Sherlock Holmes movies show scenes of thick fog, known as "pea-soupers," swirling round the gas lamps of Baker Street. These were localized problems that have largely been corrected today. The change now is that some aspects of industrialization have begun to influence the environment on a global scale. Materials are implicated in this climate change. As responsible materials engineers and scientists, we should try to understand the nature of the problem (it is not simple) and to explore what, constructively, we can do about it.

This chapter introduces the key role that materials have played in advancing technology and the dependence—*addiction* might be a better word—that it has bred. Addictions demand to be fed, and this demand, coupled with the continued growth of the human population, consumes resources at an ever-increasing rate. This situation has not, in the past, limited growth; the earth's resources are, after all, very great. But there is increasing awareness that limits *do* exist, that we are approaching some of them, and that adapting to them will not be easy.

1.2 Materials: a brief history

Materials have enabled the advance of mankind from its earliest beginnings; indeed, the ages of mankind are named after the dominant material of the day: the *Stone Age*, the *Age of Copper*, the *Bronze Age*, the *Iron Age* (see Figure 1.1). The tools and weapons of prehistory, 300,000 or more years ago, were bone and stone. Stones could be shaped into tools, particularly flint and quartz, which could be flaked to produce a cutting edge that was harder, sharper, and more durable than any other material that could be found in nature. Simple but remarkably durable structures could be built from the materials of nature: stone and mud bricks for walls, wood for beams, rush and animal skins for weather protection.

Gold, silver, and copper, the only metals that occur in native form, must have been known from the earliest time, but the realization that they were *ductile*, could be beaten to complex shape, and, once beaten, become hard, seems to have occurred around 5500 B.C. There is evidence that by 4000 B.C. man had developed technology to melt and cast these metals, allowing more intricate shapes. Native copper, however, is not abundant. Copper occurs in far greater quantities as the minerals azurite and malachite. By 3500 B.C., kiln furnaces, developed to create pottery, could reach the temperature and create the atmosphere needed to reduce these minerals, enabling the development of tools, weapons, and ornaments that we associate with the Copper Age.

But even in a worked state, copper is not all that hard. Poor hardness means poor wear resistance; copper weapons and tools were easily blunted.

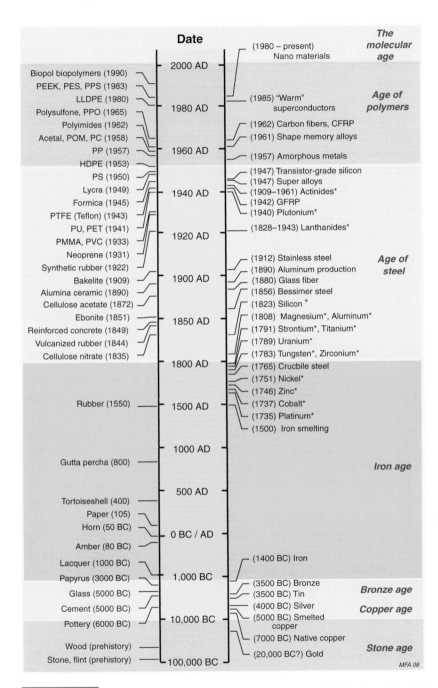

FIGURE 1.1   *The materials timeline. The scale is nonlinear, with big steps at the bottom, small ones at the top. A star (*) indicates the date at which an element was first identified. Unstarred labels give the date at which the material became of practical importance.*

Sometime around 3000 B.C. the probably accidental inclusion of a tin-based mineral, *cassiterite*, in the copper ores provided the next step in technology: the production of the alloy *bronze*, a mixture of tin and copper. Tin gives bronze a hardness that pure copper cannot match, allowing the production of superior tools and weapons. This discovery of alloying—the hardening of one metal by adding another—stimulated such significant technological advances that it, too, became the name of an era: the Bronze Age.

Obsolescence sounds like 20th-century vocabulary, but the phenomenon is as old as technology itself. The discovery, around 1450 B.C., of ways to reduce ferrous oxides to make iron, a material with greater stiffness, strength, and hardness than any other then available, rendered bronze obsolete. Metallic iron was not entirely new: tiny quantities existed as the cores of meteors that had impacted the Earth. The oxides of iron, by contrast, are widely available, particularly *hematite*, Fe_2O_3. Hematite is easily reduced by carbon, although it takes high temperatures, close to 1100°C, to do it. This temperature is insufficient to melt iron, so the material produced was a spongy mass of solid iron intermixed with slag; this was reheated and hammered to expel the slag, then forged to the desired shape.

Iron revolutionized warfare and agriculture; indeed, it was so desirable that at one time it was worth more than gold. The casting of iron, however, presented a more difficult challenge, requiring temperatures around 1600°C. Two millennia passed before, in 1500 A.D., the blast furnace was developed, enabling the widespread use of *cast iron*. Cast iron allowed structures of a new type: the great bridges, railway terminals, and civic buildings of the early 19th century are testimony to it. But it was *steel*, made possible in industrial quantities by the Bessemer process of 1856, that gave iron its dominant role in structural design that it still holds today. For the next 150 years metals dominated manufacture. The demands of the expanding aircraft industry in the 1950s, with the development of the gas turbine, shifted emphasis to the light alloys (those based on *aluminum, magnesium,* and *titanium*) and to materials that could withstand the extreme temperatures of the jet combustion chamber (*superalloys*—heavily alloyed iron and nickel-based materials). The range of application of metals expanded into other fields, particularly those of chemical, petroleum, and nuclear engineering.

The history of *polymers* is rather different. Wood, of course, is a polymeric composite, one used for construction from the earliest times. The beauty of amber (petrified resin) and of horn and tortoise shell (the polymer keratin) already attracted designers as early as 80 B.C. and continued to do so into the 19th century. (There is still, in London, a Horners' Guild, the trade association of those who worked horn and shell.) Rubber, brought to Europe in 1550, was already known and used in Mexico. Its use grew in importance in the 19th century, partly because of the wide spectrum of properties made possible by *vulcanization*—cross-linking by sulfur—giving us materials as elastic as latex and others as rigid as ebonite.

The real polymer revolution, however, had its beginnings in the early 20th century with the development of Bakelite, a phenolic, in 1909 and of synthetic butyl rubber in 1922. This was followed at midcentury by a period of rapid development of polymer science, visible as the dense group at the upper left of Figure 1.1. Almost all the polymers we use so widely today were developed in a 20-year span from 1940 to 1960, among them the bulk commodity polymers *polypropylene* (PP), *polyethylene* (PE), *polyvinyl chloride* (PVC), and *polyurethane* (PU), the combined annual tonnage of which now approaches that of steel. Designers seized on these new materials—cheap, brightly colored, and easily molded to complex shapes—to produce a spectrum of cheerfully ephemeral products. Design with polymers has since matured: they are now as important as metals in household products, automobile engineering, and, most recently, in aerospace.

The use of polymers in high-performance products requires a further step. "Pure" polymers do not have the stiffness and strength these applications demand; to provide those qualities they must be reinforced with ceramic or glass fillers and fibers, making *composites*. Composite technology is not new. Straw-reinforced mud brick (adobe) is one of the earliest of the materials of architecture, one still used today in parts of Africa and Asia. Steel-reinforced concrete—the material of shopping centers, road bridges, and apartment blocks—appeared just before 1850. Reinforcing concrete with steel gives it tensile strength where previously it had none, revolutionizing architectural design; it is now used in greater volume than any other manmade material. Reinforcing metals, already strong, took much longer, and even today metal matrix composites are few.

The period in which we now live might have been named the Polymer Age had it not coincided with yet another technical revolution, that based on silicon. Silicon was first identified as an element in 1823 but found few uses until the realization, in 1947, that, when doped with tiny levels of impurity, it could act as a rectifier. The discovery created the fields of electronics, mechatronics, and modern computer science, revolutionizing information storage, access and transmission, imaging, sensing and actuation, automation, real-time process control, and much more.

The 20th century saw other striking developments in materials technology. *Superconduction*, discovered in mercury and lead when cooled to $4.2°K$ $(-269°C)$ in 1911, remained a scientific curiosity until, in the mid-1980s, a complex oxide of barium, lanthanum, and copper was found to be superconducting at $30°K$. This triggered a search for superconductors with yet higher transition temperatures, leading, in 1987, to one that worked at the temperature of liquid nitrogen (98°K), making applications practical, though they remain few.

During the early 1990s it was realized that material behavior depended on scale and that the dependence was most evident when the scale was that of nanometers (10^{-9} m). Although the term *nanoscience* is new, technologies that use it are not. The ruby-red color of medieval stained glasses and the dia-chromic behavior of the decorative glaze known as "lustre" derive from gold nanoparticles trapped in the glass matrix. The light alloys of aerospace derive their strength from nanodispersions of intermetallic compounds. Automobile tires have, for years, been reinforced with nanoscale carbon. Modern nanotech-nology gained prominence with the discovery that carbon could form stranger structures: spherical C_{60} molecules and rod-like tubes with diameters of a few nanometers. Now, with the advance of analytical tools capable of resolving and manipulating matter at the atomic level, the potential exists to build materials the way that nature does it, atom by atom and molecule by molecule.

If we now step back and view the timeline of Figure 1.1 as a whole, clusters of activity are apparent; there is one in Roman times, one around the end of the 18th century, one in the mid-20th. What was it that trig-gered the clusters? Scientific advances, certainly. The late 18th and early 19th centuries were a time of rapid development of inorganic chemistry, particularly electrochemistry, and it was this that allowed new elements to be isolated and identified. The mid-20th century saw the birth of polymer chemistry, spawning the polymers we use today and providing key con-cepts in unraveling the behavior of the materials of nature. But there may be more to it than that. Conflict stimulates science. The first of these two periods coincides with that of the Napoleonic Wars (1796–1815), one in which technology, particularly in France, developed rapidly. And the second was that of the Second World War (1939–1945), in which technology played a greater part than in any previous conflict. One hopes that scientific prog-ress and advances in materials are possible without conflict, that the com-petitive drive of free markets is an equally strong driver of technology. It is interesting to reflect that more three quarters of all the materials scientists and engineers who have *ever* lived are alive today, and all of them are pur-suing better materials and better ways to use them. Of one thing we can be certain: there are many more advances to come.

1.3 Learned dependency: the reliance on nonrenewable materials

Now back to the main point: the environmental aspects of the way we use materials. *Use* is too weak a word; it sounds as though we have a choice: use, or perhaps not use? We don't just "use" materials, we are totally dependent on them. Over time this dependence has progressively changed from a reliance on renewable materials—the way mankind existed for thousands of years—to one that relies on materials that consume resources that cannot be replaced.

As little as 300 years ago, human activity subsisted almost entirely on renewables: stone, wood, leather, bone, natural fibers. The few nonrenewables—iron, copper, tin, zinc—were used in such small quantities that the resources from which they were drawn were, for practical purposes, inexhaustible. Then, progressively, the nature of the dependence changed (see Figure 1.2). Bit by bit nonrenewables displaced renewables until, by the end of the 20th century, our dependence on them was, as already said, almost total.

Dependence is dangerous; it is a genie in bottle. Take away something on which you depend—meaning that you can't live without it—and life becomes difficult. [Dependence exposes you to exploitation] While a resource is plentiful, market forces ensure that its price bears a relationship to the cost of its extraction. But the resources from which many materials are drawn, oil among them, are localized in just a few countries. While these compete for buyers, the price remains geared to the cost of production. But if demand exceeds supply or the producing nations reach arrangements to limit it, the genie is out of the bottle. Think, for instance, of the price of oil, which today bears little relationship to the cost of producing it.

Dependence, then, is a condition to be reckoned with. We will encounter its influence many times in subsequent chapters.

1.4 Materials and the environment

All human activity has some impact on the environment in which we live. The environment has some capacity to cope with this impact so that a certain level of impact can be absorbed without lasting damage. But it is clear that current human activities exceed this threshold with increasing frequency, diminishing the quality of the world in which we now live and threatening the well-being of future generations. Part of this impact, at least, derives from the manufacture, use, and disposal of products, and products, without exception, are made from materials.

Materials consumption in the United States now exceeds 10 tonnes per person per year. The average level of global consumption is about eight times smaller than this but is growing twice as fast. The materials (and the energy needed to make and shape them) are drawn from *natural resources*: ore bodies, mineral deposits, fossil hydrocarbons. the Earth's resources are not infinite, but until recently, they have seemed so: the demands made on them by manufacture throughout the 18th, 19th, and early 20th centuries appeared infinitesimal, the rate of new discoveries always outpacing the rate of consumption.

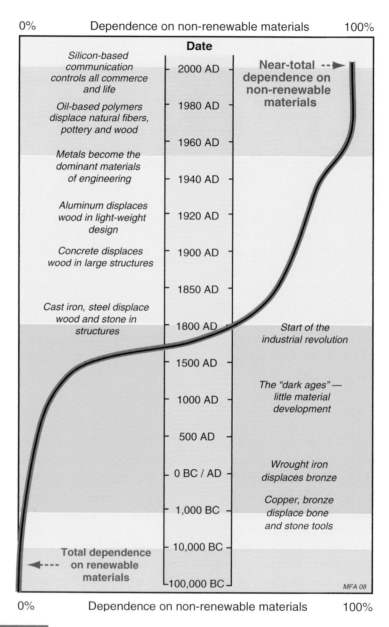

FIGURE 1.2 *The increasing dependence on nonrenewable materials over time, unimportant when they are plentiful but an emerging problem as they become scarce.*

This perception has now changed: warning flags are flying, danger signals flashing. The realization that we may be approaching certain fundamental limits seems to have surfaced with surprising suddenness, but warnings that things can't go on forever are not new. Thomas Malthus,

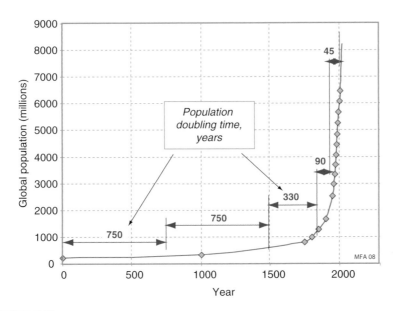

FIGURE 1.3   *Global population growth over the last 2000 years, with the doubling time marked.*

writing in 1798, foresaw the link between population growth and resource depletion, predicting gloomily that "the power of population is so superior to the power of the Earth to produce subsistence for man that premature death must in some shape or other visit the human race." Almost 200 years later, in 1972, a group of scientists known as the Club of Rome reported their modeling of the interaction of population growth, resource depletion, and pollution, concluding that "if (current trends) continue unchanged … humanity is destined to reach the natural limits of development within the next 100 years." The report generated both consternation and criticism, largely on the grounds that the modeling was oversimplified and did not allow for scientific and technological advance.

But the last decade has seen a change in thinking about this broad issue. There is a growing acceptance that, in the words of another distinguished report: "many aspects of developed societies are approaching … saturation, in the sense that things cannot go on growing much longer without reaching fundamental limits. This does not mean that growth will stop in the next decade, but that a declining rate of growth is foreseeable in the lifetime of many people now alive. In a society accustomed … to 300 years of growth, this is something quite new, and it will require considerable adjustment (WCED (1987))."

The reasons that this roadblock has sprung up so suddenly are complex, but at bottom one stands out: population growth. Examine, for a moment, Figure 1.3. It is a plot of global population over the last 2000 years. It looks

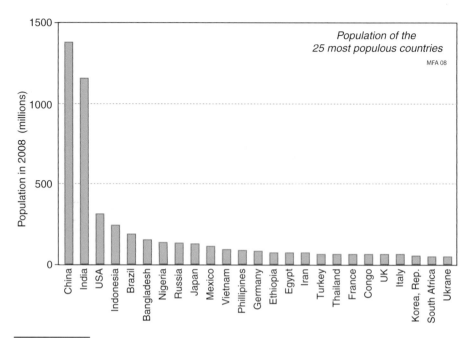

FIGURE 1.4 *The populations of the 25 most populous developed and developing countries.*

like a simple exponential growth (something we examine in more depth in Chapter 2), but it is not. Exponential growth is bad enough; it is easy to be caught out by the way it surges upward. But this is far worse. Exponential growth has a constant doubling time—if exponential, a population doubles in size at fixed, equal time intervals. The doubling times are marked on the figure. For the first 1500 years it was constant at about 750 years, but after that, starting with the industrial revolution, the doubling time halved, then halved again, then again. This behavior has been called *explosive growth*; it is harder to predict and results in a more sudden change. Malthus and the Club of Rome may have had the details wrong, but it seems they had the principle right.

Global resource depletion scales with the population and with per-capita consumption. Per-capita consumption in developed countries is stabilizing, but that in the emerging economies, as already said, is growing more quickly. Figure 1.4 shows the distribution of population in the 25 most populous nations containing between them three quarters of the global total. The first two, China and India, account for 37% of the total, and it is these two in which materials consumption is growing most rapidly.

Given all this, it makes sense to explore the ways in which materials are used in design and how this might change as environmental prerogatives become increasingly pressing. The chapters that follow explore this topic in more depth.

1.5 Summary and conclusion

Homo sapiens—that means us—differ from all other species in its competence in making things out of materials. We are not alone in the ability to *make*: termites build towers, birds build nests, beavers build dams; all creatures, in some way, make things. The difference lies in the *competence* demonstrated by humans and in their extraordinary (there can be no other word) ability to expand and adapt that competence through research and development.

The timeline of Figure 1.1 illustrates this expansion. There is a tendency to think that progress of this sort started with the Industrial Revolution, but knowledge about and development of materials have a longer and more continuous history than that. The misconception arises because of the bursts of development in the 18th, 19th, and 20th centuries, forgetting the technological developments that occurred during the great eras of the Egyptian, Greek, and Roman empires—not just to shape stone, clay, and wood and to forge and cast copper, tin, and lead, but also to find and mine the ores and to import them over great distances.

Importing tin from a remote outpost of the Roman Empire (Cornwall, England, to Rome, Italy—3300 km by sea) to satisfy the demands of the Roman State hints at an emerging materials dependence. The dependence has grown over time with the deployment of ever more manmade materials until today it is almost total. As you read this text, then, do so with the perspective that materials, our humble servants throughout history, may be evolving into our masters.

1.6 Further reading

Delmonte, J. (1985), "Origins of materials and processes", Technomic Publishing Co. ISBN 87762-420-8. *(A compendium of information about materials in engineering, documenting its history.)*

Kent, R. www.tangram.co.uk.TL-Polymer_Plastics_Timeline.html/. *(A Website devoted to the long and full history of plastics.)*

Malthus, T.R. (1798), "An essay on the principle of population", Printed for Johnson, St. Paul's Church-yard, London. www.ac.wwu.edu/~stephan/malthus/malthus. *(The originator of the proposition that population growth must ultimately be limited by resource availability.)*

Meadows, D.H., Meadows, D.L., Randers, J. and Behrens, W.W. (1972), "The limits to growth", Universe Books. *(The "Club of Rome" report that triggered the first of a sequence of debates in the 20th century on the ultimate limits imposed by resource depletion.)*

Meadows, D.H., Meadows, D.L. and Randers, J. (1992), "Beyond the limits", Earthscan. ISSN 0896-0615. *(The authors of* The Limits to Growth *use updated data and information to restate the case that continued population growth and consumption might outstrip the Earth's natural capacities.)*

Nielsen, R. (2005), "The little green handbook", Scribe Publications Pty Ltd., Carlton North. ISBN 1-920769-30-7. *(A cold-blooded presentation and analysis of hard facts about population, land and water resources, energy, and social trends.)*

Ricardo, D. (1817), "On the principles of political economy and taxation", John Murray. www.econlib.org/library/Ricardo/ricP.html. *(Ricardo, like Malthus, foresaw the problems caused by exponential growth.)*

Schmidt-Bleek, F. (1997), "How much environment does the human being need? Factor 10: the measure for an ecological economy", Deutscher Taschenbuchverlag. ISBN 3-936279-00-4. *(Both Schmidt-Bleek and von Weizsäcker, referenced below, argue that sustainable development will require a drastic reduction in material consumption.)*

Singer, C., Holmyard, E.J., Hall, A.R., Williams, T.I. and Hollister-Short, G. (Eds) (1954–2001), "A history of technology," 21 volumes, Oxford University Press. ISSN 0307-5451. *(A compilation of essays on aspects of technology, including materials.)*

Tylecoate, R.F. (1992), "A history of metallurgy", 2nd ed., The Institute of Materials. ISBN 0-904357-066. *(A total-immersion course in the history of the extraction and use of metals from 6000 BC to 1976, told by an author with forensic talent and love of detail.)*

von Weizsäcker, E., Lovins, A.B. and Lovins, L.H. (1997), "Factor four: doubling wealth, halving resource use", Earthscan Publications. ISBN 1-85383-406-8; ISBN-13: 978-1-85383406-6. *(Both von Weizsäcker and Schmidt-Bleek, referenced above, argue that sustainable development will require a drastic reduction in materials consumption.)*

WCED (1987), "Report of the World Commission on the Environment and Development", Oxford University Press, Oxford, UK. *(This document, known as the Bruntland report, sought to introduce a solid scientific underpinning to the debate on sustainability, and in doing so, highlighted the moral context of over-exploitation of resources.)*

1.7 Exercises

E.1.1. Use Google to research the history and uses of one of the following materials:

- Tin
- Glass
- Cement
- Bakelite
- Titanium
- Carbon fiber

Present the result as a short report of about 100–200 words (roughly half a page). Imagine that you are preparing it for schoolchildren. Who used the material first? Why? What is exciting about the material? Do we now depend on it, or could we, with no loss of engineering performance or great increase in cost, live without it?

E.1.2. There is international agreement that it is desirable (essential, in the view of some) to reduce global energy consumption. Producing materials from ores and feedstocks requires energy (its "embodied energy"). The following table lists the energy per kg and the annual consumption of five materials of engineering. If consumption of each material could be reduced by 10%, which material offers the greatest global energy saving? Which the least?

Material	Embodied energy MJ/kg	Annual global consumption (tonnes/year)
Steels	29	1.1×10^9
Aluminum alloys	200	3.2×10^7
Polyethylene	80	6.8×10^7
Concrete	1.2	1.5×10^{10}
Device-grade silicon	Approximately 2000	5×10^3

E.1.3. The ultimate limits of most resources are difficult to assess precisely, although estimates can be made. One resource, however, has a well-defined limit: that of usable land. The surface area of Earth is 511 million square km, or 5.11×10^{10} hectares (a hectare is 0.01 sq. km). Only a fraction of this is land, and only part of that land, is useful; the best estimate is that 1.1×10^{10} hectares of Earth's surface is biologically productive. Industrial countries require 6+ hectares of biologically productive land per head of population to support current levels of consumption. The current (2008) global population is close to 6.7 billion (6.7×10^9). What conclusions can you draw from these facts?

Resource consumption and its drivers

2.1 Introduction and synopsis

You can't understand or reach robust conclusions about human influence on the environment without a feel for the quantities involved. This chapter

The Bingham Canyon copper mine in Utah, now 1.2 km deep and 4 km across, and a Caterpiller truck that is part of the excavation equipment. (Images courtesy of Kennecott Utah Copper.)

is about their orders of magnitude. As we saw in Chapter 1, manufacturing today is addictively dependent on continuous flows of materials and energy. How big are these? A static picture—that of the values today—is a starting point, and the quantities are enormous. And of course they are *not* static. Growth is the life's blood of today's consumer-driven economies. An economy that is not growing is stagnant, sick (words used by economics correspondents for *The Times*). Business enterprises, too, seem to need to grow to survive. And all this growth causes the consumption of materials and energy to rise, or at least it has done so until now. Growth can be linear, increasing at a constant rate. It can be exponential, increasing at a rate proportional to its current size. And as we saw in Chapter 1, it can sometimes increase even faster than that.

Exponential growth plays nasty tricks. Something cute and cuddly, growing exponentially, gradually evolves into an oppressive monster. Exponential growth is characterized by a doubling time: anything growing in this way doubles its size at regular, equal intervals. Money invested in fixed-rate bonds has a doubling time, though it is usually a long one. The consumption of natural resources—minerals, energy, water—grows in a roughly exponential way: they, too, have doubling times, and some of these are short. Certain resources are so abundant that there is no concern that we are using them faster and faster; the resources from which aluminum, calcium, chlorine, hydrogen, iron, magnesium, nitrogen, oxygen, potassium, silicon, sodium, and sulfur are drawn are examples. But others are not; their ores are localized and the amount economically available is limited. Then the doubling-up nature of exponential growth becomes a concern: consumption cannot continue to double forever. And extracting and processing any material, whether plentiful or scarce, uses energy—lots of energy—and it too is a resource under duress.

This situation sounds alarming, and many alarming statements have been made about it. But consider this: the exhaustion time for reserves of the ores of copper, in 1930, was estimated to be 30 years; today (2008) the exhaustion time of copper reserves is calculated as … 30 years. There is obviously something going on here besides exponential growth. This chapter explores it.

2.2 Resource consumption

Materials. Speaking globally, we consume roughly 10 billion (10^{10}) tonnes of engineering materials per year, an average of 1.5 tonnes per person, though it is not distributed like that. Figure 2.1 gives a perspective: it is a bar chart showing consumption of the materials used in the greatest

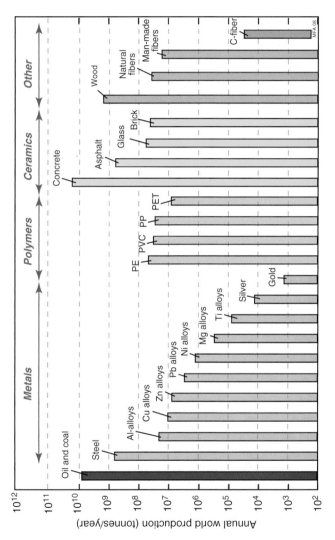

FIGURE 2.1   The annual world production of 23 materials on which industrialized society depends. The scale is logarithmic.

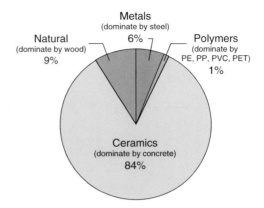

FIGURE 2.2 *A pie chart of materials usage (tonnes) by family. Ceramics dominate because of the enormous annual consumption of concrete.*

quantities. The chart has some interesting messages. On the extreme left, for calibration, are hydrocarbon fuels—oil and coal—of which we currently consume a colossal 9 billion (9×10^9) tonnes per year. Next, moving to the right, are metals. The scale is logarithmic, making it appear that the consumption of steel (the first metal) is only a little greater than that of aluminum (the next); in reality, the consumption of steel exceeds, by a factor of 10, that of all other metals combined. Steel may lack the high-tech image that attaches to materials like titanium, carbon-fiber enforced composites, and (most recently) nanomaterials, but make no mistake, its versatility, strength, toughness, low cost, and wide availability are unmatched.

Polymers come next. Fifty years ago their consumption was tiny; today the consumption of the commodity polymers polyethylene (PE), polyvinyl chloride (PVC), polypropylene (PP), and polyethylene-terephthalate (PET) exceeds that of any metal except steel. The really big ones, though, are the materials of the construction industry. Steel is one of these, but the consumption of wood for construction purposes exceeds even that of steel, even measured in tonnes per year (as in Figure 2.1), and since it is a factor of 10 lighter, if measured in m³/year, wood totally eclipses steel. Bigger still is the consumption of concrete, which exceeds that of all other materials combined. The other big ones are asphalt (roads) and glass. The last column of all illustrates things to come: it shows today's consumption of carbon fiber. Just 20 years ago this material would not have crept onto the bottom of this chart. Today its consumption is approaching that of titanium and is growing much more quickly.

Figure 2.2 presents some of this information in another way, as a pie chart showing the tonnage-fraction of each family of material: metals, polymers,

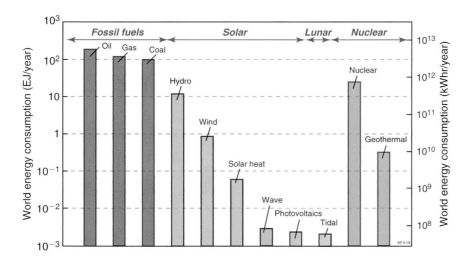

FIGURE 2.3 *The annual world consumption of energy by source. The units on the left are exajoules (10^{18} J); those on the left are the more familiar kWhr. (Data from Nielsen, 2005.)*

ceramics, and natural. The logarithmic scale of Figure 2.1 can be deceptive; the true magnitude of the consumption of concrete only emerges when plotted in this second way. This is important when we come to consider the impact of materials on the environment, since impact scales with the quantity consumed.

Energy. Although this book is about materials, energy appears throughout; it is inseparable from the making of materials, their manufacture, their use, and their disposal. The SI unit of energy is the joule (J), but because it is very small, we generally use kJ (10^3 J), MJ (10^6 J), or GJ (10^9 J) as the unit. Power is joules/sec, or watts (W), but it too is small, so we usually end up with kW, MW, or GW. The everyday unit of energy is the kW.hr, one kW drawn for 3600 seconds, so 1 kW.hr = 3.6 MJ.

Where does energy come from? There are really only four sources:

- The sun, which drives the winds, wave, hydro, photochemical, and photoelectric phenomena
- The moon, which drives the tides
- Nuclear decay of unstable elements inherited from the creation of the Earth, providing geothermal heat and nuclear power
- Hydrocarbon fuels, the sun's energy in fossilized form

All, ultimately, are finite, but the time scale for the exhaustion of the first three is so large that it is safe to regard them as infinite.

How much energy do we use? Energy is measured in joules, but when we speak of world consumption the unit is the exajoule, symbol EJ, a billion

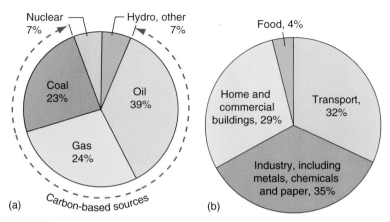

FIGURE 2.4 *World energy consumption (a) by source and (b) by use. The nonrenewable carbon-based fuels oil, gas, and coal account for 86% of the total.*

billion (10^{18}) joules. The value today (2008) is about 500 EJ, and, of course, it is rising. It comes from the sources shown in Figure 2.3. Fossil fuels dominate the picture, providing about 86% of the total (Figure 2.4a). Nuclear gives about 7%. Hydro, wind, wave, solar heat, and photovoltaics add up to just another 7%. These sun-driven energy pools are enormous, but unlike fossil and nuclear fuels, which are concentrated, they are distributed. The problem is that of harvesting the energy, something we return to in later chapters.

Where does the energy go? Most of it goes into three big sectors: transportation, buildings (heating, cooling, lighting), and industry, including the production of materials (Figure 2.4b).

Figures 2.3 and 2.4 focus on the *sources of energy*. To be useful you have to do something with energy, and that, almost always, involves *energy conversion*. Figure 2.5 suggests the possible conversion paths, not all of them practical perhaps, but there in principle. Almost always, conversion carries energy "losses." Energy, the first law of thermodynamics tells us, is conserved, so it can't really be lost. What happens is that in almost all energy conversion, some energy is converted to heat. *High-grade heat* is heat at high temperature, as in the burning gas of a power station or the burning fuel of the combustion chamber of an IC engine or gas turbine; it can be used to do work. *Low-grade heat* is heat at low temperature, and for reasons we get to in a moment, it is not nearly so useful; indeed, most of it is simply allowed to escape, and in this sense it is "lost." Energy conversion generally has low-grade heat as a by-product. The exception is the conversion electrical to thermal energy; it has an efficiency η of 100%, meaning that all

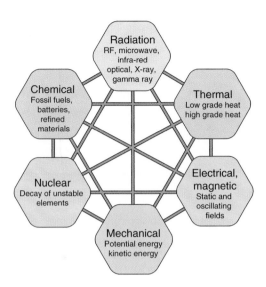

FIGURE 2.5 *The six types of energy. Each can be transformed into the others, as indicated by the linking lines, but at conversion efficiencies that differ greatly.*

the iR energy dissipated by the electrical current ends up as heat. But that is where it ends. The refining of metals from their oxide, sulfide, or other ores, for instance, involves the conversion of thermal or electrical energy into chemical energy—the energy that could (in principle) be recovered by allowing the metal to reoxidize or resulfidize. The recovered energy, of course, is a lot less than the energy it took to do the refining; the conversion efficiency is low.

The conversion efficiencies of the many paths in Figure 2.5 differ greatly even when "ideal," meaning that the conversion is as good as the laws of physics allow. Conversion of thermal to mechanical energy, for instance, is ultimately limited by the Carnot efficiency, η_c:

$$\eta_c = 1 - \frac{T_{out}}{T_{in}}$$

where T_{in} is the temperature of the steam or hot gas entering the heat engine and T_{out} is the temperature at which it exits. Plotted, this looks like Figure 2.6. The maximum input temperature T_{in} is limited by the materials of the turbine blades and discs (around 650°C for steam turbines, as much as 1400°C for gas turbines) and by the exhaust temperature T_{out} (at least 150°C), giving a theoretical maximum efficiency of about 75%.

The real conversion efficiencies of heat engines are much less than this. Steam engines of the early 19th century had efficiencies of perhaps 2%. The

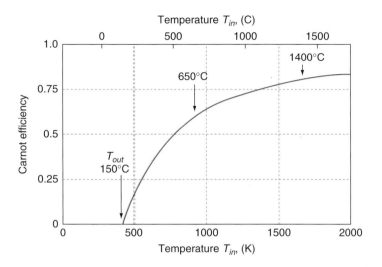

FIGURE 2.6 *Carnot efficiency of heat engines as a function of temperature T_{in}, assuming $T_{out} = 150°C$.*

first electricity-generating plant built in Holborn, London, in 1882 had an efficiency of just 6%. A modern steam-turbine plant today achieves 38%, although the national average in a country like the United States or the United Kingdom is less than this (around 35%, a value we use later). A plant driven by a gas turbine can reach 50%, a consequence of a higher T_{in} and lower T_{out}. Table 2.1 lists practical efficiencies of other energy conversion processes. Many are low. Thus a more useful measure of efficiency η is simply the ratio:

$$\eta = \frac{Power\ out}{Power\ in}$$

where *Power* is energy, in any form, per unit time. This is the definition used in the rest of the book.

Using energy to extract and refine materials is an example of an energy conversion process. The energy per unit mass consumed in making a material from its ores and feedstock is known as its *embodied energy*. The term is a little misleading; only part of this energy is really "embodied" in the sense that you could get it back. That bit is the free energy difference between the refined material and the ore from which it came. Take a metal—say, iron—as an example. Iron is made by the reduction of the oxide *hematite*, Fe_2O_3, in a blast furnace. Thermodynamics dictates that the minimum energy needed to do this is 6.1 MJ/kg; it is the free energy of oxidation of iron to this particular oxide. This energy could, in principle, be recovered by reoxidizing the

Table 2.1	Some approximate efficiency factors for energy conversion		
Energy conversion path	**Efficiency, direct conversion (%)**	**Efficiency relative to oil equivalence (%)**	**Associated carbon (kg CO_2 per useful MJ)**
Fossil fuel to thermal, enclosed system	100	100	0.07
Fossil fuel to thermal, vented system	65–75	70	0.10
Fossil fuel to electric	33–39	35	0.20
Fossil fuel to mechanical, steam turbine	28–42	40	0.17
Fossil fuel to mechanical, gas turbine	46–50	48	0.15
Electric to thermal	100	35	0.20
Electric to mechanical, electric motors	85–93	31	0.23
Electric to chemical, lead-acid battery	80–85	29	0.24
Electric to chemical, advanced battery	85–90	31	0.23
Electric to em radiation, incandescent lamp	15–20	6	1.17
Electric to em radiation, LED	80–85	30	0.23
Light to electric, solar cell	10–20	—	0

iron under controlled conditions (it is truly "embodied"). But the measured energy to make iron—the quantity that is called its *embodied energy*—is three times larger, at about 18 MJ/kg, a conversion efficiency of 33%. Where has the rest gone? Mostly as heat lost in the gases of the blast furnace. Similar losses reduce the efficiency of all material synthesis and refining.

Water. Manufacture draws on another resource, that of water. Water is a renewable resource, but renewable only at the rate that the ecosystem allows. Increasing demand is now putting supply under growing pressure: the worldwide demand for water has tripled over the past 50 years. Forecasts suggest that water might soon become as important an issue as oil is today, with more than half of humanity short of water by 2050. Agriculture is the largest consumer worldwide, taking about 65% of all fresh water, but in industrialized countries it is industry that is the big user.

The water demands of materials and manufacture are measured directly as factory inputs and outputs. The unit used here is liters of water per kg of material produced, l/kg (or, equivalently, kg/kg, since a liter of water weighs 1 kg). The range for engineering materials extends from 10 l/kg to over 1000 l/kg. What is water used for? In the production of steel, as an example, water is used in the extraction of the minerals (iron ore, limestone, and

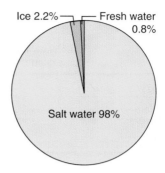

Ice 2.2% — ⌐ Fresh water 0.8%

Salt water 98%

FIGURE 2.7 *The global distribution of water. Only a tiny fraction is accessible as fresh water.*

fossil fuels), for material conditioning (dust suppression), pollution control (scrubbers to clean up waste gases), and cooling material and equipment.

We ignore saltwater usage. Seawater is extremely abundant, making up 97% of the water on Earth. Accessible fresh water accounts for less than 1%; the remainder is fresh water locked up as ice in glaciers and icecaps (see Figure 2.7). The water consumption for the growth of natural materials requires a distinction between those that are irrigated and those that are not. Plant life that is the source of materials such as wood, bamboo, cork, and paper is not, as a general rule, irrigated, whereas plants used in the production of some thermoplastics (cellulose polymers, polyhydroxyalkanoates, polylactides, starch-based thermoplastics) and for cattle feedstock (for leather) generally require irrigation. For this reason it is usual to split water usage into two parts: commercial water usage and total water usage. For most materials these two values are the same, but whereas the total water usage in the growing of trees and plants includes nonirrigation water, the commercial water usage is just that used for irrigation. The data in the data sheets in Chapter 12 of this book are for commercial water usage.

The provision of energy, too, uses water for cooling cycles (with loss by evaporation) and for dust suppression and washing. Table 2.2 lists water usage per MJ of delivered energy for electricity, both produced and distributed via a public grid system, and electricity produced industrially. (Industrial electricity generation is more efficient because the hot gases produced can be used in other processes, whereas they are simply vented in electricity production for the grid.)

2.3 Exponential growth and doubling times

A modern industrialized state is extremely complex, heavily dependent on a steady supply of raw materials. Most materials are being produced at a

Table 2.2	The water demands of energy
Energy source	**Liters of water per MJ**
Grid electricity	24
Industrial electricity	11
Energy direct from coal	0.35
Energy direct from oil	0.3

rate that is growing exponentially with time, at least approximately, driven by increasing global population and standards of living. So we should look first at exponential growth and its consequences.

If the current rate of production of a material is P tonnes per year and this increases by a fixed fraction $r\%$ every year, then

$$\frac{dP}{dt} = \frac{r}{100}P \qquad (2.1)$$

Integrating over time t gives

$$P = P_0 \ exp\left\{\frac{r(t - t_0)}{100}\right\} \qquad (2.2)$$

where P_0 is the production rate at time $t = t_0$. Figure 2.8, left, shows how P grows at an accelerating rate with time. Taking logs of this equation gives

$$\log_e\left(\frac{P}{P_0}\right) = 2.3 \log_{10}\left(\frac{P}{P_0}\right) = \frac{r}{100}(t - t_0)$$

so a plot of $log_{10}(P)$ against time t, as on the right of Figure 2.8, is *linear* with a slope of $r/230$.

Figure 2.9 shows the production of three metals over the past 100 years, plotted, at top, on linear scales and, at bottom, on semilog scales, exactly as in the previous figure. The broken lines show the slopes corresponding to growth at $r = 2\%$, 5%, and 10% per year. Copper and zinc production has grown at a consistent 3% per year over this period. Aluminum, initially, grew at nearly 7% per year but has now settled back to about 4%.

Exponential growth is characterized by a *doubling-time* t_D over which production doubles in size. Setting $P/P_0 = 2$ in Equation 2.2 gives:

$$t_D = \frac{100}{r} \log_e(2) \approx \frac{70}{r} \qquad (2.3)$$

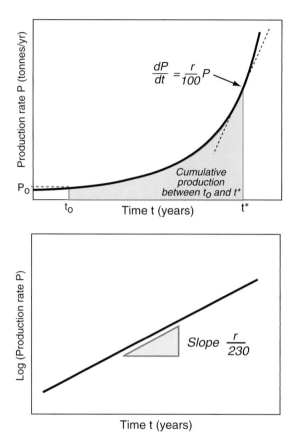

FIGURE 2.8 *Exponential growth. Production* P *doubles in a time* $t_d \approx 70/r$, *where* r *% per year is the annual growth rate.*

The cumulative production Q_t between times t_0 and t^* is found by integrating Equation 2.2 over time, giving:

$$Q_t = \int_{t_0}^{t^*} P\,dt = \frac{100P_0}{r}\left(exp\left\{\frac{r(t^* - t_0)}{100}\right\} - 1\right) \qquad (2.4)$$

This result illustrates the most striking feature of exponential growth: at a global growth rate of just 3% per year we will mine, process, and dispose of more "stuff" in the next 25 years than in the entire 300 years since the start of the Industrial Revolution. An alarming thought, pursued further in Exercise E.2.5 at the end of this chapter.

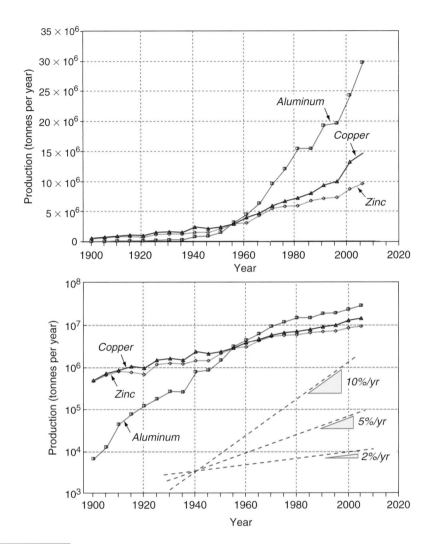

FIGURE 2.9 *The growth of production of three metals over a 100-year interval, plotted on linear and semilogarithmic scales.*

2.4 Reserves, the resource base, and resource life

The materials on which industry depends are drawn, very largely, from the Earth's reserves of minerals. A *mineral reserve, R,* is defined as that part of a known mineral deposit that can be extracted legally and economically at the time it is determined. It is natural to assume that reserves describe the *total* quantity of minerals present in the ground that is accessible and, once used, is gone forever, but this is wrong. In reality, reserves are an economic construct, which grow and shrink under varying economic, technical, and legal conditions. Improved extraction technology can enlarge them, but environmental legislation or changing political climate may make them shrink. Demand stimulates prospecting, with the consequence that reserves tend to grow in line with consumption. The world reserves of lead, for instance, are three times larger today than they were in 1970; the annual production has increased by a similar factor.

The *resource base* (or just *resource*) of a mineral is the real total, and it is much larger. It includes not only the current reserves but also all usable deposits that might be revealed by future prospecting and that, by various extrapolation techniques, can be estimated. It includes, too, known and unknown deposits that cannot be mined profitably now but which—due to higher prices, better technology, or improved transportation—may become available in the future. Although the resource base is much larger than the reserves, much of it is inaccessible using today's technology, and its evaluation is subject to great uncertainty.

The distinction between reserves and resources is illustrated by Figure 2.10. It has axes showing the *degree of certainty* with which the mineral is known to exist and the *ore grade*, a measure of the richness

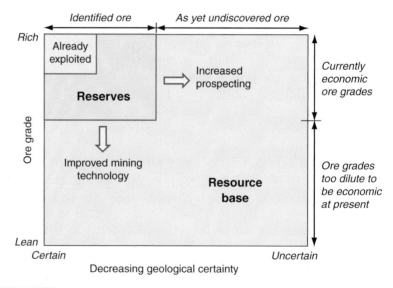

FIGURE 2.10 *The distinction between* reserves *and the* resource base. *The resource base is fixed. The reserves are the part of this base that has been discovered and established as economically viable for extraction.*

of the ore and, indirectly, of the ease and cost of extracting it. The largest rectangle represents the resource base. The smaller green-shaded rectangle represents the reserves, of which a small part, shaded in gray, has been depleted by past exploitation. The reserves are extended downward by improved mining technology or by an increase in price (because this allows leaner ores to be mined profitably), and they are extended to the right by prospecting. A number of factors cause the reassignment of resources into and out of the *reserve* classification. They include:

- *Commodity price.* As metal prices rise, it becomes profitable to mine lower-grade ore.

- *Improved technology.* New extractive methods can increase the economically workable ore grade.

- *Production costs.* Rising fuel or labor costs can make deposits uneconomic.

- *Legislation.* Tightening or loosening environmental laws can increase or decrease production costs or enable or deprive access to exploitable deposits.

- *Depletion.* Mining consumes reserves; prospecting enlarges them. A rate of production that exceeds that of discovery—the sign that there are problems ahead—causes the reserve to shrink.

Thus the reserves from which materials derive are elastic. It is nonetheless important to have a figure for their current value so that mining companies can assess their assets and governments can ensure availability of materials critical to the economy. There are procedures for estimating the current size of reserves; the U.S. Geological Survey, for example, does so annually. The data sheets in Chapter 12 of this book list the most recent values available at the time of writing.

If you know for a material the size of its reserve and of its *annual world production* (also listed in the data sheets), it might seem that estimates of *resource criticality* could be got by dividing the one by the other. That, too, is wrong. The argument follows.

Resource criticality: time to exhaustion. The availability of any commodity depends on the balance between supply and demand. The *material supply chain*, sketched in Figure 2.11, has a supply side from which material flows into stock. The demand side draws material out, depleting the stock. In a free market, market forces keep the two in long-term balance, though there can be short-term imbalance because increased demand causes the

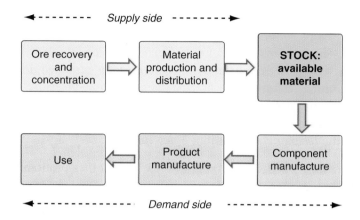

FIGURE 2.11 *The supply chain for a material. If the market works efficiently, the supply side and the demand side remain in balance. Scarcity arises when demand exceeds supply.*

supply side to respond by increasing production to compensate. Material scarcity appears when the supply chain fails to respond in this way. Failure can have many origins. The most obvious is that the resource becomes so depleted that it can no longer be exploited economically.

This has led to attempts to predict resource life. If you have D dollars in the bank and you spend it at the rate of S dollars per year without topping it up, you can expect a letter from your bank manager about D/S years from now. The equivalent when speaking of reserves is the exhaustion time, the so-called *static index of exhaustion*, $t_{ex,s}$:

$$t_{ex,s} = \frac{R}{P} \qquad (2.5)$$

where R (tonnes) is the reserve and P (tonnes per year) is the production rate. But this ignores the growth. As we have seen, the production rate P is not constant but generally increases with time at a rate r per year. Allowing for this means that the reserves will be consumed in the time:

$$t_{ex,d} = \frac{100}{r} \ln\left\{ \frac{rR}{100P_o} + 1 \right\} \qquad (2.6)$$

known as the *dynamic index*.

Figure 2.12 shows the static and dynamic index for copper over the last 70 years. The static index has hovered around 40 years for the whole of that time; the dynamic index has done the same at 30. As a prediction of resource exhaustion, neither one inspires confidence. Do either of these

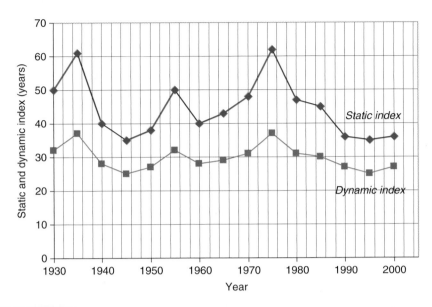

FIGURE 2.12 *The static and dynamic index for copper over the last 70 years.*

make sense? The message they convey is *not* that we will soon run out of copper; it is that a comfort zone for the value of the index exists and that it is around 30 years. Only when it falls below this value is there sufficient incentive to prospect for more. We need to approach the problem of resource criticality in another way.

Market balance and breakdown. Thus far we have assumed (without saying so) that the market works efficiently, keeping the supply side and the demand side in balance. When it does, increased demand is met by increased prospecting and improved technology. The price of material then reflects the true cost of its extraction. But what happens when market forces don't work? That is a thought that troubles both businesses and government.

Some resources are widely distributed, but others are not. For these the ore bodies that are rich enough to be worked economically are located in just a few countries; this is called *supply chain concentration*. Then political unrest, economic upheaval, rebellion, or forced change of government in one of these or its neighbors (through which the ore must be transported) disrupts supply in ways to which the market cannot immediately respond. And there is the potential, too, for the few supplier nations to reach agreement to limit supply, thereby driving up prices, a process known as *cartel action*. The market cannot immediately respond, because it takes time to set up new extraction facilities and establish new supply chains.

If the supplier that stocks your favorite wine closes, you seek another supplier. If that supplier, already under pressure, has only limited stock, you could try to buy it all and store it, creating a stockpile. If the supplier refuses, you have to consider reducing your consumption or—horrors—finding a substitute, keeping your favorite wine for special occasions. That, too, is the reaction of the demand side of the market: seek other suppliers, stockpile, explore substitutes for all but the most demanding applications, and (something that does not work with wine) increase recycling.

Can scarcity be predicted? That caused by failure of the supply chain becomes less likely when the chain is diverse, meaning that the ore deposits are widely distributed and there are many producers and distributors. For some materials the reverse is true; then attempts to control price by limiting supply become more likely. Scarcity caused by depletion can be countered by developing extraction technologies to deal with leaner ores and by increased recycling.

With this background, let us return to the question of criticality.

More realistic indicators of criticality. The argument here gets a little more complex. Figure 2.13 helps set it out. The resource base—the big rectangle of Figure 2.10—is, of course, finite. The broken curve of Figure 2.13 shows schematically how the reserves—the exploitable part of the resource base—at first grows as prospecting and improved extraction technology reveal more. Its exploitation (the red curve) begins to eat it away, but initially the

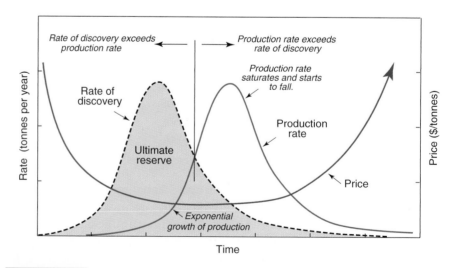

FIGURE 2.13 *The evolution of the rates of discovery of a resource (broken curve) and its exploitation, the rate of production (red curve) and the price (blue curve).*

rate of discovery reassuringly exceeds that of exploitation and no alarm bells ring. But there comes a point at which the finding of new deposits and further improved technology become more difficult and the reserves, though still growing, grow at a rate that falls behind that of production. This decline in discovery rate is followed, with a characteristic time lag, by a decline in production.

This progression is reflected in the price of minerals. When first discovered and exploited, a mineral is expensive (blue line in Figure 2.13). As extractive technology improves and prospecting unveils richer deposits, the price falls. Though reserves remain large, the price, when corrected for inflation, settles to a plateau. But then, with about the same time lag as that for fall in production, depletion makes itself felt and the price climbs. The crossing point is significant: once past this point, the reserves start to shrink—they are being used faster than they are being topped up. There are indicators of criticality here that can be taken seriously:

- The rate of growth of discovery falls below the rate of growth of production.
- The production rate curve peaks and starts to decline.
- The minimum economic ore grade falls.
- The price starts to rise sustainable.

In reality, the curves are not smooth. Figure 2.13 is a schematic; Figure 2.14 is real. It shows oil discoveries, production, and price corrected to the value of the dollar in 2000 over time since 1900.

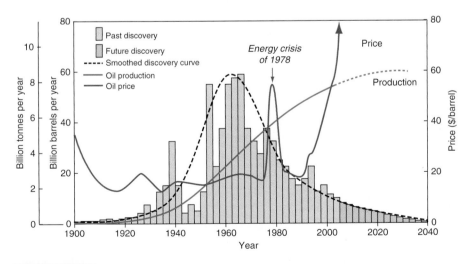

FIGURE 2.14   *The evolution of the rates of discovery, the rate of production, and the price of oil.*

Here some of the fluctuations are included. Major discoveries do not occur every year, giving the lumpy profile of the green blocks. Anticipated discovery between now and 2040 are the orange blocks. Another big lump or two is not impossible—the competing claims by Russia, Norway, Canada, and the United States for territorial rights in the Arctic suggest that there may be large deposits there. Ignoring this for the moment, a smooth curve (the black broken line) has been sketched in. Growth in production (the red one) is much smoother; the demand has grown steadily over the past 100 years. There have been some fluctuations (not shown), but these are small. That of price (the blue curve on the scale at right) is much more erratic. Oil price does not reflect the actual cost of extraction but the degree to which production is controlled to maintain price. As reserves are depleted, those of the greatest consumers—the United States and Western Europe—become exhausted, creating dependence on imports and vulnerability to production quotas; the spike in the 1970s and that of today are examples. The fluctuations of the real data are distracting, but if these are smoothed out, say, using a 10-year moving average of the data, the evolution is that of the schematic. We are past the crossover point of the discovery and production-rate curves. It seems likely that at least part of the recent rise in price is not a temporary fluctuation but is here to stay. Price increases as large as this, however, open new doors. Oil sands and offshore Arctic oil, for example, have until now been largely inaccessible because of the cost of exploitation. They now become economically attractive, extending the tail of the estimated production curve in Figure 2.14.

The crossover points of the discovery and production curves for most minerals from which engineering materials are drawn have not yet been reached. For many it is still far off. But at least we now have measures that mean something.

2.5 Summary and conclusion

Growing global population and prosperity increase the demand for energy and materials. The growth in demand is approximately exponential, meaning that consumption grows at a rate that is proportional to its current value; for most materials it is between 3% and 6% per year. Exponential growth has a number of consequences. One is that consumption doubles every $70/r$ years, where r is the growth rate in percent per year. It also means that the total amount consumed (the integral of the consumption over time) also doubles in the same time interval.

Most materials are drawn from the minerals of the Earth's land masses and oceans. The *resource base* from which they are drawn is large, but it is not infinite. Its magnitude is not easy to estimate so that at any point in time only a fraction of it, the *reserve*, is established as accessible and economically viable. Market forces, when they operate properly, ensure that reserves remain adequate, keep supply in line with demand, and ensure that prices remain stable and fair.

The balance between supply and demand can, however, be disrupted. Depletion of the resource base causes scarcity, driving prices up. Reserves that are localized can become vulnerable to cartel action or cut off by local political unrest. Market disruption can be economically damaging, so foreseeing and anticipating it is necessary if stability is to be maintained. Economic forecasting of this sort is based on the tracking of discovery and production rates of material resources and the identification of their sources, flagging those for which a single source dominates world markets as vulnerable.

2.6 Further reading

Alonso, E., Gregory, J., Field, F. and Kirchain, R. (2007), "Material availability and the supply chain: risks, effects and responses", Environmental Science and Technology, Vol. 41, pp. 6649–6656. (*An informative analysis of the causes of instability in material price and availability.*)

Battery technology: http://en.wikipedia.org/wiki/Rechargeable_battery _Comparison_of_battery_types. (*Information on battery efficiencies.*)

Chapman, P. F., Roberts, F. (1983), "Metal resources and energy", Butterworth's Monographs in Materials, Butterworth and Co., ISBN 0-408-10801-0. (*A monograph that analyses resource issues, with particular focus on energy and metals.*)

McKelvey, V. E. (1973), Technology Review, March/April p. 13. (*The original presentation of the McKelvey diagram.*)

Shell Petroleum (2007), "How the energy industry works", Silverstone Communications Ltd., ISBN 978-0-9555409-0-5. (*Useful background on energy sources and efficiency.*)

USGS (2007), "Mineral Information, Mineral yearbook *and* Mineral commodity summary", http://minerals.usgs.gov/minerals/pubs/commodity/. (*The gold standard information source for global and regional material production, updated annually.*)

Wolfe, J. A. (1984), "Mineral resources: a world review", Chapman & Hall, ISBN 0-4122-5190-6. (*A survey of the mineral wealth of the world, both for metals and nonmetals, describing their extraction and the economic importance of each.*)

2.7 Exercises

E.2.1 Explain the distinction between reserves and the resource base.

E.2.2 The world consumption rate of CFRP is rising at 8% per year. How long does it take to double?

E.2.3 Derive the dynamic index:

$$t_{ex,d} = \frac{100}{r} \log_e \left\{ \frac{rR}{100 P_o} + 1 \right\}$$

starting with Equation 2.2 of the text.

E.2.4 A total of 5 million cars were sold in China in 2007; in 2008 the sale was 6.6 million. What is the annual growth rate, expressed as % per year? If there were 15 million cars already on Chinese roads by the end of 2007 and this growth rate continues, how many cars will there be in 2020, assuming that the number that are removed from the roads in this time interval can be neglected?

E.2.5 Prove the statement made in the text that, "At a global growth rate of just 3% per year, we will mine, process, and dispose of more 'stuff' in the next 25 years than in the entire 300 years since the start of the Industrial Revolution.'"

E.2.6 Understanding reserves: copper. The following table lists the world production and reported reserves of copper over the last 20 years.

Year	Price (US$/kg)	World production (millions of tonnes/Year)	Reserves (millions of tonnes)	Reserves/world production (years)
1995	2.93	9.8	310	31.6
1996	2.25	10.7	310	29.0
1997	2.27	11.3	320	28.3
1998	1.65	12.2	340	27.9
1999	1.56	12.6	340	27.0
2000	1.81	13.2	340	25.8
2001	1.67	13.7	340	24.8
2002	1.59	13.4	440	32.8
2003	1.78	13.9	470	33.8
2004	2.86	14.6	470	32.2
2005	3.7	14.9	470	31.5
2006	6.81	15.3	480	31.4

- Examine trends (plot price, production, and reserves against time): what do you conclude?
- Tabulate the reserves/world production to give the static index of exhaustion. What does the result suggest about reserves?

E.2.7 The following table shows the production rate and the reserves of five metals over a period of 10 years. What has been the growth rate of production? What is that of the reserves? What conclusions can you draw about the criticality of the material?

Metal	Year	Production rate, tonnes/year	Reserves, ronnes
Platinum	2005	217	71×10^3
	1995	145	56×10^3
Nickel	2005	1.49×10^6	64×10^6
	1995	1.04×10^6	47×10^6
Lead	2005	3.27×10^6	67×10^6
	1995	2.71×10^6	55×10^6
Copper	2005	15.0×10^6	480×10^6
	1995	10.0×10^6	310×10^6
Cobalt	2005	57.5×10^3	7×10^6
	1995	22.1×10^3	4×10^6

E.2.8 Tabulate the annual world production in tonnes/year and the densities (kg/m^3) of carbon steel, PE, soft wood, and concrete. (You will find the data in data sheets for these materials in the second part of this book; use an average of the ranges given in the data sheets.) Calculate, for each, the annual world production measured in m^3/year. How does the ranking change?

E.2.9 The price of cobalt, copper, and nickel have fluctuated wildly in the past decade. Those of aluminum, magnesium, and iron have remained much more stable. Why? Research this topic by examining uses (is it one of high value-added goods?) and the localization of the producing mines. The USGS Website listed under Further Reading is a good starting point.

E.2.10 The production of zinc over the period 1992–2006 increased at a rate of 3.1% per year. The reserves, over the same period, increased 3.5%. What conclusions about the criticality of zinc supply can you draw from these figures?

E.2.11 The production of platinum, vital for catalysts and catalytic converters, has risen from 145 to 217 tonnes per year over the last 10 years. The ores are highly localized in South Africa, Russia and Canada. The reserves have risen from 56,000 to 71,000 tonnes in the same time interval. Would you classify platinum as a critical material? Base your judgement on the relative growth rates of production and reserves, and on the dynamic index (equation 2.6) calculated using 2005 data.

E.2.12 Global water consumption has tripled in the last 50 years. What is the growth rate, $r\%$, in consumption assuming exponential growth? By what factor will water consumption increase between 2008 and 2050?

E.2.13 Plot the *Annual world production* of metals against their *Price*, using mean values from the data sheets in Chapter 12 of this book. What trend is visible?

Exercises using CES Eco Level 2

E.2.14 Use CES to plot the *Annual world production* of materials against their *Price*. What trend is visible?

E.2.15 Make a plot of apparent resource life for materials (the reserves in tonnes divided by the annual production rate in tonnes per year), using the "Advanced" facility in CES to plot the ratio. Which metals have the longest apparent resource life? Why are these apparent lives not a reliable measure of the true life of the mineral?

CHAPTER 3

The materials life cycle

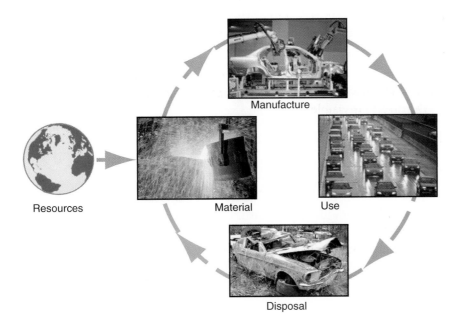

Manufacture

Resources

Material

Use

Disposal

3.1 Introduction and synopsis

The materials of engineering have a life cycle. They are created from ores and feedstock. These are manufactured into products that are distributed and used. Like us, products have a finite life, at the end of which they become scrap. The materials they contain, however, are still there; some (unlike us) can be resurrected and enter a second life as recycled content in a new product.

Life-cycle assessment (LCA) traces this progression, documenting the resources consumed and the emissions excreted during each phase of life. The output is a sort of biography, documenting where the materials have been, what they have done, and the consequences for their surroundings.

Image of casting courtesy of Skillspace; image of car making courtesy of U.S. Department of Energy EERE program; image of cars courtesy of Reuters.com; image of junk car courtesy of Junkyards.com.

LCA can take more than one form: a full LCA that scrutinizes every aspect of life (arduous and expensive in time and money), a briefer "character sketch" painting, an approximate (but still useful) portrait; or something in between. All, however, should observe certain boundary conditions. They take a bit of explaining.

This chapter is about the life cycle of materials and its assessment: how LCAs work, their precision (or lack of it), the difficulties of implementing them, and ways that these can be bypassed to guide material choice. It ends by introducing a strategy that is developed in the chapters that follow. The chapter includes an appendix describing currently available LCA software and the usual Further Reading and Exercises.

3.2 The material life cycle

The idea of a *life cycle* has its roots in the biological sciences. Living organisms are born; they develop, mature, grow old, and, ultimately, die. The progression is inherent in the organism—all follow the same path—but the way the organism develops on the way and its behavior and influence depend on its interaction with its *environment*—here, the natural environment. Life-cycle studies explore and track the interaction of organisms with their environment.

The life-cycle idea has since been adapted and applied in other fields: in the social sciences (the interaction of individuals with their social environment), in the management of technology (the study of innovation in the business environment), and in product design (the interaction of products with the natural, social, and business environments). Concern about resource depletion (the Club of Rome Report, already described), the oil crisis of the early 1970s, followed by the first evidence of carbon-induced global warming, focused attention on the last of these: the life cycle of manufactured products and their interaction, above all, with the natural environment. Products are made of materials; materials are their flesh and bones, so to speak, and these are central to the interaction. The study of product and associated materials life cycles involves assessing the environmental impacts associated with the full life of products, from the extraction of raw materials to their return to the ecosphere as waste—from birth to death, or (if you prefer) from cradle to grave. That means tracking materials through life. So let us explore that idea.

Figure 3.1 is a sketch of the material's life cycle. Ore, feedstock, and energy, drawn from the planet's natural resources on the left, are processed to give materials. These are manufactured into products that are distributed,

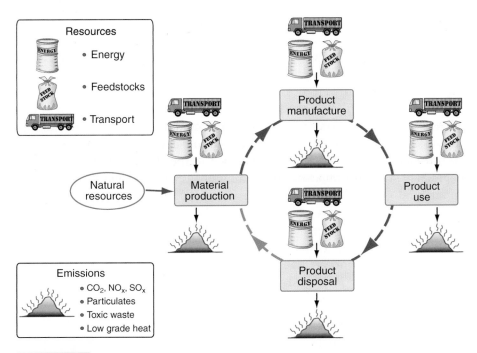

FIGURE 3.1 *The material life cycle. Ore and feedstock are mined and processed to yield a material. This material is manufactured into a product that is used, and at the end of its life, it is discarded, recycled, or, less commonly, refurbished and reused. Energy and materials are consumed in each phase, generating waste heat and solid, liquid, and gaseous emissions.*

sold, and used. Products have a useful life, at the end of which they are discarded, a fraction of the materials they contain perhaps entering a recycling loop, the rest committed to incineration or landfill.

Energy and materials are consumed at each point in this cycle, depleting natural resources. Consumption brings an associated penalty of carbon dioxide (CO_2), oxides of sulfur (SO_x), and of nitrogen (NO_x), and other emissions in the form of low-grade heat and gaseous, liquid, and solid waste. In low concentrations, most of these emissions are harmless, but as their concentrations build, they become damaging. The problem, simply put, is that the sum of these unwanted by-products now often exceeds the capacity of the environment to absorb them. For some the damage is local and the creator of the emissions accepts the responsibility and cost of containing and remediating it (the environmental cost is said to be *internalized*). For others the damage is global and the creator of the emissions is not held directly responsible, so the environmental cost becomes a burden on society as a whole (it is *externalized*). The study of resource consumption, emissions, and their impacts is called *life-cycle assessment* (LCA).

3.3 Life-cycle assessment: details and difficulties

Formal methods for LCA first emerged in a series of meetings organized by the Society for Environmental Toxicology and Chemistry (SETAC), of which the most significant were held in 1991 and 1993. This led, from 1997 on, to a set of standards for conducting an LCA, issued by the International Standards Organization (ISO 14040 and its subsections 14041, 14042, and 14043). These prescribe procedures for "defining goal and scope of the assessment, compiling an inventory of relevant inputs and outputs of a product system; evaluating the potential impacts associated with those inputs and outputs; interpreting the results of the inventory analysis and impact assessment phases in relation to the objectives of the study." The study must (according to the ISO standards) examine energy and material flows in raw material acquisition, processing and manufacture, distribution and storage (transport, refrigeration, and so forth), use, maintenance and repair, recycling options, and waste management. There is a lot here and there is more to come. A summary in plainer English might help:

- *Goals and scope.* Why do the assessment? What is the subject, and which bit(s) of its life are assessed?

- *Inventory compilation.* What resources are consumed, what emissions excreted?

- *Impact assessment.* What do these do to the environment— particularly, what bad things do they do?

- *Interpretation.* What do the results mean and what is to be done about them?

We look now at what each of these questions involves.

Goals and scope. Why do the study? To what purpose? And where does such a study begin and end? Figure 3.2 shows the four phases of life, each seen as a self-contained unit, with notional "gates," marked as $\otimes$, through which inputs pass and outputs emerge. If you were the manager of the man- ufacturing unit, as an example, your goal might be to compile an assessment of your plant, ignoring the other three phases of life because they are beyond your control. This is known as a *gate-to-gate study*, the scope limited to the activity inside the box labelled System Boundary A. There is a tendency for the individual life phases to minimize energy use, material waste, and internalized emission costs spontaneously because it saves money to do so. But this action by one phase may have the result of raising resource consumption and emissions of the others. For example, if minimizing

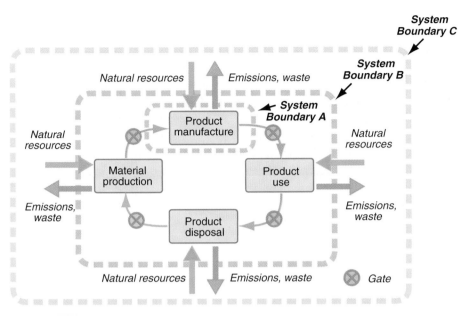

FIGURE 3.2 *LCA system boundaries with the flows of resources and emissions across them.*

the manufacturing energy and material costs for a car results in a heavier vehicle and one harder to disassemble at end of life, the gains made in one phase have caused losses in the other two. Put briefly: the individual life phases tend to be self-optimizing; the system as a whole does not. We return to this topic in Chapters 8 and 9, where the necessary trade-off methods are developed.

If the broader goal is to assess the resource consumption and emissions of the product over its entire life, the boundary must enclose all four phases (System Boundary B). The scope becomes that of product birth to product death, including, at birth, the ores and feedstock that are drawn from the Earth's resources and, at death, the consequences of disposal.

Some LCA proponents see a still more ambitious goal and grander scope (System Boundary C). If ores and feedstock are included (as they are within System Boundary B), why not the energy and material flows required to make the equipment used to mine or resource them? And what about the flows to make the equipment that made *them*? Here an injection of common sense is needed. Setting the boundaries at infinity gets us nowhere. Equipment-making facilities make equipment for other purposes too, and this gives a dilution effect: the more remote they are, the smaller the fraction of their resources and emissions that is directly linked to the product being assessed. The standards are vague on how to deal with this point, merely instructing that the system boundary "shall be determined," leaving the scope of the assessment as a subjective decision. A practical way forward is to include only the primary flows directly entering the materials, manufacture use, and disposal of the product, excluding the secondary ones required to make the primary possible. We shall follow this route in the audits of Chapter 7.

Inventory compilation. Setting the boundaries is the first step. The second is data collection: amassing an inventory of the resource flows passing into the system and the emissions passing out, per unit of useful output. But how is this data to be measured? Per kilogram of final product? Yes, if the product is sold and used by weight. Per m^3 of final product? Yes, if it is sold by volume. But few products are sold and used in this way. More usually it is neither of these, but per *unit of function*, a point we will return to in later chapters. The function of a container for a soft drink (a soda bottle, a plastic water bottle, a beer can) is to contain fluid. The bottle maker might measure resource flows per bottle, but if it is concerned with environmental or economic consequences of its entire life, it is the eco-impact or cost *per unit volume of fluid contained* that is the proper measure. Refrigerators provide a cooled environment and maintain it over time. The maker might measure flows per fridge, but the logical measure from a life-cycle standpoint is the resource consumption *per unit of cooled volume per unit time* (cold m^3/year, for instance).

We will find that the functional unit of a resource entering one phase is not the same as those leaving it. There is nothing subtle about this, it's just to make accounting easier. Thus the flow of materials leaving Phase 1 of life and entering Phase 2 *are* traded by weight, so the functional unit here is per unit weight: the embodied energy of copper, for instance, is listed in the data sheet as 68–74 MJ/kg. The output of manufacture is a component or product; here kilograms or product units may be used. It is in the use phase that the function becomes important and here the real measure—that per unit of function—becomes the logical one.

The inventory analysis, then, assesses resource consumption and emissions per functional unit. It is also necessary to decide on the level of detail—the granularity—of the assessment. It doesn't make sense to include every nut, bolt, and rivet. But where should the cut-off come? One proposal is to include the components that make up 95% of the weight of the product, but this is risky; electronics, for instance, don't weigh much, but the resources and emissions associated with their manufacture can be large, a point we return to in Chapter 6.

Figure 3.3 is a schematic of the start of an inventory analysis—the identification of the main resources and emissions—here for a washing

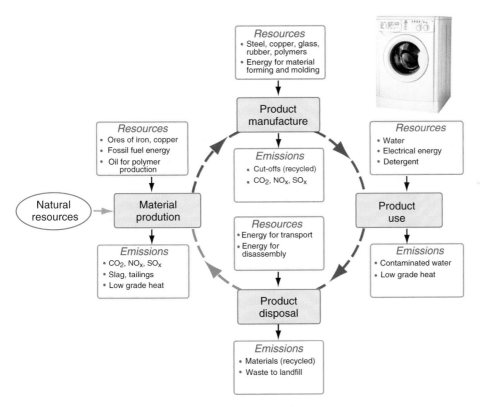

FIGURE 3.3 *The principle resource emissions associated with the life cycle of a washing machine.*

machine. Most of the parts are made of steel, copper, plastics, and rubber. Both materials production and product manufacture require carbon-based energy with associated emissions of CO_2, NO_x, SO_x, and low-grade heat. The use phase consumes water as well as energy, with contaminated water as an emission. Disposal creates burdens typical of any large appliance.

Impact assessment. The inventory, once assembled, lists resource consumption and emissions, but they are not equally malignant; some have a greater impact than others. Impact categories include *resource depletion, global warming potential, ozone depletion, acidification, eutrophication,*[1] *human toxicity*, and more. Each impact is calculated by multiplying the quantity of each inventory item by an *impact assessment factor*—a measure of how profoundly a given inventory type contributes to each impact

---

[1] *Eutrophication* is the over-enrichment of a body of water with nutrients—phosphates, nitrates—resulting in excessive growth of organisms and depletion of oxygen concentration.

Table 3.1	Example global warming potential impact assessment factors	
Gas		**Impact assessment factor**
Carbon dioxide, CO_2		1
Carbon monoxide, CO		1.6
Methane, CH_4		21
Di-nitrous monoxide, N_2O		256

category. Table 3.1 lists some examples of that for assessing global warming potential. The overall impact contribution of a product to each category is found by multiplying the quantity emitted by the appropriate impact assessment factor and summing the contributions of all the components of the product for all four phases of life.

Interpretation. The final questions: what do these inventory and impact values mean? What should be done to reduce their damaging qualities? The ISO standard requires answers to these questions but gives little guidance about how to reach them. All this makes a full LCA a time-consuming matter requiring experts. Expert time is expensive. A full LCA is not something to embark on lightly.

The output and its precision. Figure 3.4 shows part of the output of a partial LCA—here, one for the production of aluminum cans (it stops at the

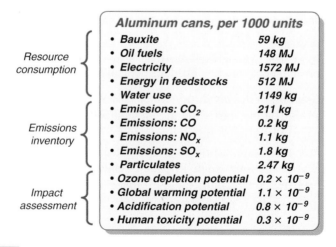

FIGURE 3.4 *Typical LCA output showing three categories: Resource consumption, emission inventory, and impact assessment. (Data in part from Bousted, 2007.)*

exit gate of the manufacturing plant, so this is a *cradle-to-gate*, not a *cradle-to-grave* study). The functional unit is "per 1000 cans." There are three blocks of data: one an inventory of resources, one of emissions, and one of impacts—here only some of them.

Despite the formalism that attaches to LCA methods, the results are subject to considerable uncertainty. *Resource* and *energy* inputs can be monitored in a straightforward and reasonably precise way. The *emissions* rely more heavily on sophisticated monitoring equipment; few are known to better than $\pm 10\%$. Assessments of *impacts* depend on values for the marginal effect of each emission on each impact category; many of these have much greater uncertainties.

And there are two further difficulties, both serious. First, what is a designer supposed to do with these numbers? The designer, seeking to cope with the many interdependent decisions that any design involves, inevitably finds it hard to know how best to use data of this type. How are CO_2 and SO_x emissions to be balanced against resource depletion, energy consumption, global warming potential, or human toxicity? And second, how is it to be paid for? A full LCA takes days or weeks. Does the result justify this considerable investment? LCA has value as a *product assessment tool*, but it is not a *design tool*.

Aggregated measures: eco-indicators. The first of these difficulties has led to efforts to condense the LCA output into a single measure, or *eco-indicator*. To do this, four steps are necessary, as shown in Figure 3.5. The first is that of *classification* of the data listed in Figure 3.4 according to the impact each causes (global warming, ozone depletion, acidification, etc.). The second step is that of *normalization* to remove the units (of which there are several in the LCA report) and reduce them to a common scale (0–100, for instance). The third step is that of *weighting* to reflect the perceived seriousness of each impact; thus global warming might be seen as more serious than resource depletion, giving it a larger weight. In the final step, the weighted, normalized measures are *summed* to give the indicator.[2] Eco-indicators have found most use in condensing eco-information for the first phase of life, that of material production. Values for materials, when available, are included in the data sheets in Chapter 12 of this book.

The use of a single-valued indicator is criticized by some. The grounds for criticism are that there is no agreement on normalization or weighting factors, that the method is opaque since the indicator value has no simple physical significance, and that defending design decisions based on a

---

[2] Details can be found in EPS (1993), Idemat (1997), EDIP (1998), and Wenzel et al. (1997) Eco-indicator (1999).

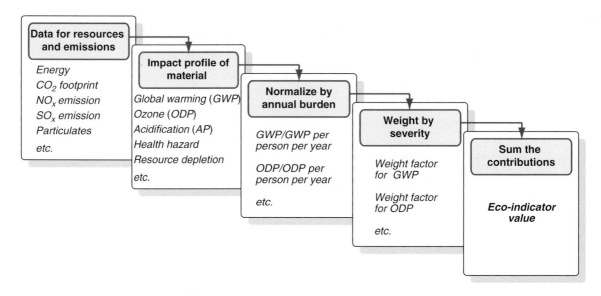

FIGURE 3.5 *The steps in calculating an eco-indicator. Difficulty arises in Step 3; there is no agreement on how to choose the weight factors.*

measurable quantity such as *energy consumption* or *CO_2 release to atmosphere* carry more conviction than doing so with an indicator.

3.4 Streamlined LCA

Emerging legislation imposes ever-increasing demands on manufacturers for eco-accountability. The EU Directive 2005/32/EC on Energy Using Products (EuPs), for example, requires that manufacturers of EuPs must demonstrate "that they have considered the use of energy in their products as it relates to materials, manufacture, packaging, transport, use and end of life." This sounds horribly like a requirement that a full LCA be conducted on each one of a manufacturer's products; because many manufacturers have thousands of products, the expense in terms of both money and time would be prohibitive.

The complexity of an LCA makes it, for many purposes, unworkable. This perception has stimulated two lines of development: simplified, or "streamlined," methods of assessment that focus on the most significant inputs, neglecting those perceived to be secondary; and software-based tools that ease the task of conducting an LCA. Software solutions are documented in the appendix to this chapter. We turn now to streamlining.

	Material	Manufacture	Transport	Use	Disposal
Material resources	$M_{1,1}$	$M_{1,2}$			
Energy use	$M_{2,1}$	etc.			
Global warming					
Human health					
The biosphere					

FIGURE 3.6 *An example of a streamlined LCA matrix.*

The matrix method. The detail required for a full LCA excludes its use as a design tool; by the time the necessary detail is known, the design is too far advanced to allow radical change. *Streamlined LCA* attempts to overcome this problem by basing the study on a reduced and simplified inventory of resources, accepting a degree of approximation while retaining enough precision to guide decision making. One approach is to simplify while still attempting a *quantitative* analysis—one using numbers. This method is developed in later chapters of this book and illustrated with case studies. The other—one developed by Graedel[3] and others and used in various forms by a number of industries—is *qualitative*. The matrix of Figure 3.6 shows the idea. The life phases appear as the column headers, the impacts as the row headers. An integer between 0 (highest impact) and 4 (least impact) is assigned to each matrix element M_{ij}, based on experience guided by checklists, surveys, or protocols.[4] The overall *Environmentally Responsible Product Rating,* R_{erp}, is the sum of the matrix elements:

$$R_{erp} = \sum_i \sum_j M_{ij} \qquad (3.1)$$

Alternative designs are ranked by this rating.

[3] Graedel (1998); Todd and Curran (1999). See Further reading at the end of the chapter.

[4] Graedel lists an extensive protocol in the appendix to the book listed under Further reading.

FIGURE 3.7 *It is now standard practice to report official fuel economy figures for cars (e.g., Combined: 6–11 liter/100 km, CO_2 emissions: 158–276 g/km) and energy ratings for appliances (e.g., 330 kWhr/year, efficiency rating: A).*

There are many variants of this approach, differing in the impact categories of the rows and the life (or other) categories of the columns. The method has the merit that it is flexible, is easily adapted to a variety of products, carries a low overhead in time and effort, and, in the hands of practitioners of great experience, can take into account the subtleties of emissions and their impacts. It has the drawback that it relies heavily on judgment. It is not a tool to put in the hands of a novice. Is there an alternative?

One resource, one emission. There is, as yet, no consensus on a metric for eco-impact of product life that is both workable and able to guide design. On one point, however, there is a degree of international agreement:[5] a commitment to a progressive reduction in carbon emissions, generally interpreted as meaning CO_2. At the national level the focus is more on reducing energy consumption, but since this and CO_2 production are closely related, reducing one generally reduces the other. Thus there is a certain logic in basing design decisions on energy consumption or CO_2 generation. They carry more conviction than the use of a more obscure indicator, as evidenced by the now-standard reporting of both energy efficiency and the CO_2 emissions of cars and the energy rating and ranking of appliances (Figure 3.7) dealing with the use phase of life. To justify this further, we digress briefly to glance at the IPCC report of 2007.

The 2007 IPCC report. The Intergovernmental Panel on Climate Change (IPCC), an international study group set up by the World Meteorological Organization and the United Nations Environmental Panel, has published a

---

[5] The Kyoto Protocol of 1997 and subsequent treaties and protocols, detailed in Chapter 5.

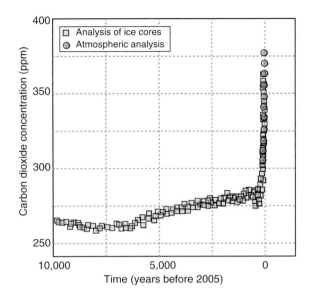

FIGURE 3.8 *Atmospheric concentration of CO_2 over the last 10,000 years measured from ice cores and atmospheric samples. Redrawn from the IPCC report of 2007.*

series of reports on the effect of industrial activity on the biosphere and the human environment. The most recent of these (IPCC, 2007) is of such significance that familiarity with it is a prerequisite for thinking about sustainability and the environment. Briefly, the conclusions it reaches are these:

- The average air, ocean, and land-surface temperatures of the planet are rising. The increase is causing widespread melting of snow and ice cover, rising sea levels, and changes of climate.

- Climate change, measured, for instance, by the average of these temperatures, affects natural ecosystems, agriculture, animal husbandry, and human environments. As little as 1°C rise in average global temperature can have a significant effect on all of them. A rise of 5° would be catastrophic.

- The global atmospheric concentration[6] of CO_2 has increased at an accelerating rate since the start of the industrial revolution (around 1750) and is now at its highest level for the past 600,000 years. Most of the increase has occurred between 1950 and the present day (see Figure 3.8).

[6] Throughout this book carbon release to the atmosphere is measured in kg of CO_2. One kg of elemental carbon is equivalent to 3.6 kg of CO_2.

■ Increasingly accurate geophysical measurement allow the history of temperature and atmospheric carbon to be tracked, and increasingly precise meteorological models allow scenario exploration and prediction of future trends in both. The two together establish beyond all reasonable doubt that the climate-temperature rise is caused by greenhouse gases, and that, though there are causes, it is manmade CO_2 that is principally responsible. The potential consequences of further rise in climate temperature, details in the IPCC report, give cause for concern, to which we return in Chapters 10 and 11.

The point, then, is that, of the many emissions associated with industrial activity, it is CO_2 that is of greatest current concern. It is global in its impact, causing harm both to the nations that generate most of it and those that do not. It is closely related to the consumption of fossil fuels, themselves a diminishing resource. And if the IPCC report is to be taken seriously, the urgency to cut carbon emissions is great. At this stage in structuring our thinking about materials and the environment, taking energy consumption and the release of atmospheric CO_2 as metrics is a logical simplification.

3.5 The strategy for eco-selection of materials

The need, as we've already said, is for an assessment strategy that addresses current concerns and combines acceptable cost burden with sufficient precision to guide decision making. The strategy should be flexible enough to accommodate future refinement and simple enough to allow rapid "What if?" exploration of alternatives. To achieve this goal, it is necessary to strip off much of the detail, multiple targeting, and complexity of method that makes standard LCA techniques so cumbersome. The approach developed here has three components.

Adopt simple metrics of environmental stress. The preceding discussion points to the use of energy or of a CO_2 footprint as the logical choices. The two are related and are understood by the public at large. Energy has the merit that it is the easiest to monitor, can be measured with relative precision, and, with appropriate precautions, can, when needed, be used as a proxy for CO_2.

Distinguish the phases of life. Figure 3.9 suggests the breakdown, assigning a fraction of the total life-energy demands of a product to material creation, product manufacture, transport, and product use and disposal. Product disposal can take many different forms, some carrying an energy

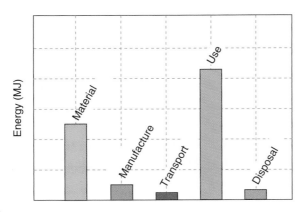

FIGURE 3.9 *Breakdown of energy into that associated with each life phase.*

penalty, some allowing energy recycling or recovery. Because of this ambiguity, disposal has a chapter (Chapter 4) to itself.

When this distinction is made, it is frequently found that one of the phases of Figure 3.1 dominates the picture. Figure 3.10 presents the evidence. The upper row shows an approximate energy breakdown for three classes of energy-using products: a civil aircraft, a family car, and an appliance. For all three the use phase consumes more energy than the sum of all the other phases. The lower row shows products that still require energy during the use phase of life, but not as intensively as those of the upper row. For these, the embodied energies of the materials of which they are made make the largest contribution.

Two conclusions can be drawn. The first: one phase frequently dominates, accounting for 60% or more of the energy—often much more. If large energy savings are to be achieved, it is the dominant phase that becomes the first target, since it is here that a given fractional reduction makes the biggest contribution. The second: when differences are as great as those of Figure 3.10, great precision is not necessary; modest changes to the input data leave the ranking unchanged. It is the nature of people who measure things to want to do so with precision, and precision must be the ultimate goal. But it is possible to move forward without it; precise judgments can be drawn from imprecise data. Chapter 7 explains how the breakdowns are made and gives examples.

Base the subsequent action on the energy or carbon breakdown. Figure 3.11 suggests how the strategy can be implemented. If material production is the dominant phase, the logical way forward is to choose materials

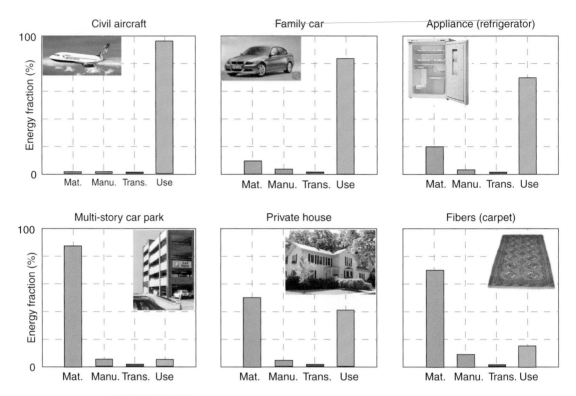

FIGURE 3.10 *Approximate values for the energy consumed at each phase of Figure 3.1 for a range of products (data from References 5 and 6). The disposal phase is not shown because there are many alternatives for each product.*

with low embodied energy and to minimize the amounts that are used. If manufacture is an important energy-using phase of life, reducing processing energies becomes the prime target. If transport makes a large contribution, seeking a more efficient transport mode or reducing distance becomes the first priority. When the use phase dominates, the strategy is to minimize mass (if the product is part of a system that moves), to increase thermal efficiency (if a thermal or thermomechanical system), or to reduce electrical losses (if an electromechanical system). In general the best material choice to minimize one phase will not be the one that minimizes the others, requiring trade-off methods to guide the choice.

Implementation requires tools. Two sets are needed: one to perform the eco-audit sketched in the upper part of Figure 3.11, the other to enable the analysis and selection of the lower part. The first, the eco-audit tool, is described in Chapter 7. The second, that of optimized selection, is the subject of Chapters 8 and 9. Tools require data. Chapter 12 of this book contains

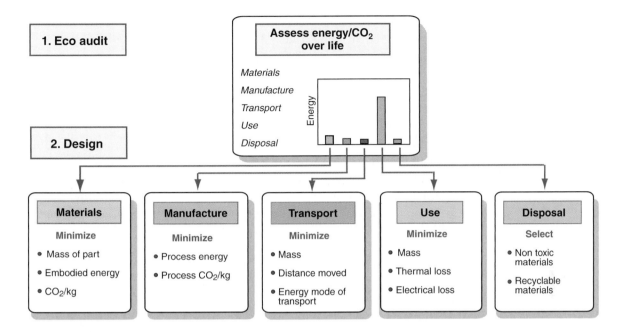

FIGURE 3.11   *Rational approaches to the ecodesign of products start with an analysis of the phase of life to be targeted. Its results guide redesign and materials selection to minimize environmental impact. The disposal phase, shown here as part of the overall strategy, is not included in the current version of the tool.*

data sheets for materials, documenting their engineering and ecoproperties.[7] The engineering properties are familiar. The ecoproperties are less so; Chapter 6 explores them.

3.6 Summary and conclusion

Products, like organisms, have a life, during the course of which they interact with their environment. Their environment is also ours; if the interaction is a damaging one, it diminishes the quality of life of all who share it.

Life-cycle assessment is the study and analysis of this interaction, quantifying the resources consumed and the waste emitted. It is holistic, spanning the entire life from the creation of the materials through the manufacture of the product, its use, and its subsequent disposal. Although

---

[7] The data sheets are a subset of those contained in the CES (2009) software, which also implements both the tools described here.

standards now prescribe procedures for doing this, they remain vague, allowing a degree of subjectivity. Implementing them requires skill and experience, and it requires access to much detail, making a full LCA an expensive and time-consuming proposition and one that, in its present form, delivers output that is not helpful to designers.

No surprise. The technique of LCA is relatively new and is still evolving. The framework prescribed by the ISO is not well adapted to current needs. The way forward is to adopt a less precise but much simpler approach, limiting the assessment to key aspects of the interaction to streamline it. The matrix method, of which there are many variants, assigns a ranking for each impact category in each phase of life, summing the rankings to get an eco-impact rating. Another approach, better adapted to guiding material choice, is to limit the impact categories to one resource—energy—and one emission—CO_2—auditing designs or products for their demands on both. Provided the resolution of the audit is sufficient to draw meaningful conclusions, the results can guide the strategy for material selection or substitution.

The chapter ends by summarizing a strategy, one that is developed more fully in the chapters that follow. The appendix contains a brief review of current software to help with life-cycle assessment.

3.7 Further reading

Aggregain, The Waste and Resources Action Program (WRAP), www.wrap.org.UK, 2007, ISBN 1-84405-268-0. (*Data and an Excel-based tool to calculate energy and carbon footprint of recycled road-bed materials.*)

Allwood, J.M., Laursen, S.E., de Rodriguez, C.M. and Bocken, N.M.P. (2006), "Well dressed? The present and future sustainability of clothing and textiles in the United Kingdom", University of Cambridge, Institute for Manufacturing. ISBN 1-902546-52-0. (*An analysis of the energy and environmental impact associated with the clothing industry.*)

Baxter Sustainability Report, 2007, http://sustainability.baxter.com/product_responsibility/materials_use.html. (*Analysis of end of life.*)

Boustead Model 5, Boustead Consulting, 2007, www.boustead-consulting.co.uk. (*An established life-cycle assessment tool.*)

Eco-indicator (1999) PRé Consultants, Printerweg 18, 3821 AD Amersfoort, The Netherlands (www.pre.nl/eco-indicator 99/eco-indicator_99.htm)

EPS, The EPS enviro-accounting method: an application of environmental accounting principles for evaluation and valuation in product design, Report B1080, IVL Swedish Environmental Research Institute, by B. Steen and S. O. Ryding, 1992.

EU Directive on Energy Using Products, Directive 2005/32/EC of the European Parliament and of the Council of July 6, 2005, establishing a framework for the setting of ecodesign requirements for energy-using products, and amending

Council Directive 92/42/EEC and Directives 96/57/EC and 2000/55/EC of the European Parliament and of the Council, 1995. (*One of several EU Directives relating to the role of materials in product design.*)

GaBi, PE International, 2008, www.gabi-software.com. (*GaBi is a software tool for product assessment to comply with European legislation.*)

Goedkoop, M., Effting, S., and M. Collignon, The eco-indicator 99: a damage-oriented method for life cycle impact assessment, manual for designers, April 14, 2000, www.pre.nl. (*An introduction to eco-indicators, a technique for rolling all the damaging aspects of material production into a single number.*)

Graedel, T.E. and Allenby, B.R. (2003), "Industrial ecology", 2nd edition, Prentice Hall. (*An established treatise on industrial ecology.*)

Graedel, T.E. (1998) Streamlined life-cycle assessment, Prentice Hall, ISBN 0-13-607425-1. (*Graedel is the father of streamlined LCA methods. The first half of this book introduces LCA methods and their difficulties. The second half develops his streamlined method with case studies and exercises. The appendix details protocols for informing assessment decision matrices.*)

GREET, Argonne National Laboratory and the U.S. Department of Transport, 2007, www.transportation.anl.gov/. (*Software for analyzing vehicle energy use and emissions.*)

Guidice, F., La Rosa, G. and Risitano, A. (2006), "Product design for the environment", CRC/Taylor and Francis. ISBN 0-8493-2722-9. (*A well-balanced review of current thinking on ecodesign.*)

Heijungs, R. (Ed.) (1992), "Environmental life-cycle assessment of products: background and guide", Netherlands Agency for Energy and Environment.

Idemat Software version 1.0.1 (1998), "Faculty of Industrial Design Engineering", Delft University of Technology. (*An LCA tool developed by the University of Delft, Holland.*)

ISO 14040, (1998), "Environmental management: life-cycle assessment", Principles and framework.

ISO 14041, (1998), "Goal and scope definition and inventory analysis".

ISO 14042, (2000), "Life-cycle impact assessment".

ISO 14043 (2000), "Life-cycle interpretation", International Organization for Standardization. (*The set of standards defining procedures for life-cycle assessment and its interpretation.*)

Kyoto Protocol, United Nations, Framework Convention on Climate Change, Document FCCC/CP1997/7/ADD.1, 1997, http://cop5.unfccc.de. (*An international treaty to reduce the emissions of gases that, through the greenhouse effect, cause climate change.*)

MEEUP Methodology Report, final, VHK, (2005) www.pre.nl/EUP/. (*A report by the Dutch consultancy VHK commissioned by the European Union, detailing their implementation of an LCA tool designed to meet the EU Energy-Using Products directive.*)

MIPS, The Wuppertal Institute for Climate, Environment and Energy, 2008, www.wupperinst.org/en/projects/topics_online/mips/index.html. (*MIPS software uses an elementary measure to estimate the environmental impacts caused by a product or service.*)

National Academy of Engineering and National Academy of Sciences (1997), "The industrial green game: implications for environmental design and management", National Academy Press. ISBN 978-0309-0529-48. (*A monograph describing best practices that are being used by a variety of industries in several countries to integrate environmental considerations in decision making.*)

SETAC (1991), A technical framework for life-cycle assessment, Fava, J.A., Denison, R., Jones, B., Curran, M.A., Vignon, B., Selke, S., and Barnum, J., (Eds.), Society of Environmental Toxicology and Chemistry. (*The meeting at which the term Life Cycle Assessment was first coined.*)

SETAC (1993), Guidelines for life-cycle assessment: a code of practice, Consoli, F., Fava, J.A., Denison, R., Dickson, K., Kohin, T., and Vigon, B., (Eds.), Society of Environmental Toxicology and Chemistry. (*The first formal definition of procedures for conducting an LCA.*)

Todd, J.A., and Curran, M.A. Streamlined life-cycle assessment: a final report from the SETAC North America streamlined LCA workshop, Society of Environmental Toxicology and Chemistry, 1999. (*One of the early moves toward streamlined LCA.*)

3.8 Appendix: software for LCA

The most common uses of life-cycle assessment are for product improvement ("How can I make my products greener?"), support of strategic choices ("Is this or that the greener development path?"), benchmarking ("How do our products compare?"), and for communication ("Our products are the greenest"). Most of the software tools designed to help with this task use ISO 14040 to 14043 as a prescription. In doing so they commit themselves to a process of considerable complexity.[8] There is no compulsion to follow this route, and some do not. Some of these are aimed at specific product sectors (vehicle design, building materials, paper making), others at the early stages of product design and these, of necessity, are simpler in their structure. Two, at least, have education as their target. So there is quite a spectrum, 11 of which are listed in Table 3.2. Some of these programs are free, some can be bought, and others are available only through the services of a consultant—an understandable precaution, given the complexity of using them properly.

> *SimaPro (2008).* SimaPro 7.1 is a widely used tool to collect, analyze, and monitor the environmental performance of products and services developed by Pré Consultants in the Netherlands. Life cycles

---

[8] Pré Consultants estimate that the time needed to perform a "screening" LCA is about eight days and for a full LCA is about 22 days.

Table 3.2	LCA and LCA-related software
Tool name	**Provider**
SimaPro	Pré Consultants (www.pre.nl)
Boustead model 5	Boustead Consultants (www.boustead-consulting.co.uk)
TEAM (EcoBilan)	PriceWaterhouseCooper (www.ecobalance.com)
GaBi	PE International (www.gabi-software.com)
MEEUP method	VHK, Delft, Netherlands (www.pre.nl/EUP/)
GREET	US Department of Transportation (www.transportation.anl.gov/)
MIPS	Wuppertal Institute (www.wupperinst.org)
CES Eco '09	Granta Design, Cambridge UK (www.grantadesign.com)
Aggregain	WRAP (www.aggregain.org.uk)
KCL-ECO 3.0	KCL Finland (www.kcl.fi)
Eiloca	Carnegie Mellon Green Design Institute, USA (www.eiolca.net)

can be analyzed in a systematic way, following the ISO 14040 series recommendations. There is an educational version. A free demo is available from the Pré Website.

Boustead Model 5 (2007). The Boustead Model is a tool for life-cycle inventory calculations, broadly following the ISO 14040 series recommendations. Ian Boustead, the author of the software, has many years of experience in life-cycle assessment, working with European polymer suppliers.

TEAM (2008). TEAM is Ecobilan's life-cycle assessment software. It allows the user to build and use a large database and to model systems associated with products and processes following the ISO 14040 series of standards.

GaBi (2008). GaBi 4, developed by PE International, is a sophisticated tool for product assessment to comply with European legislation. It has facilities for analyzing cost, environment, social and technical criteria, and optimization of processes. A demo is available.

MEEUP method (2005). The Dutch Methodology for Ecodesign of Energy-using Products (MEEUP) is a response to the EU directive on energy-using products (the EuP Directive) described in Chapter 5. It is a tool for the analysis of products—mostly appliances—that use energy, following the ISO 14040 series of guidelines.

GREET (2007). The Greenhouse Gasses, Regulated Emissions, and Energy Use in Transportation Model (GREET) is a free spreadsheet running in Microsoft Excel and developed by Argonne National Laboratory for the U.S. Department of Transportation. There are two versions: one for fuel-cycle analysis and one for vehicle-cycle analysis. They deal with specific emissions, not with impacts and weighted combinations. For a given vehicle and fuel system, the model calculates energy consumption, emissions of CO_2-equivalent greenhouse gases—primarily carbon dioxide (CO_2), methane (CH_4), and nitrous oxide (N_2O)—and six criteria pollutants: volatile organic compounds (VOCs), carbon monoxide (CO), nitrogen oxide (NO_x), particulate matter with size smaller than 10 microns (PM10), particulate matter with size smaller than 2.5 microns (PM2.5), and sulfur oxides (SO_x).

MIPS (2008). MIPS stands for Material Input per Service Unit. MIPS is an elementary measure to estimate the environmental impacts caused by a product or service. The full life cycle from cradle to cradle (extraction, production, use, waste/recycling) is considered. It allows the environmental implications of products, processes, and services that need to be assessed and compared. It enables material intensity analysis both at the micro level (focusing on specific products and services) and at the macro level (focusing on national economies).

CES Eco (2009). Granta Design specializes in materials information management software. One of their products, CES Eco, is a widely used tool for teaching engineering students about the selection and use of materials and processes. It includes modules that implement the eco-audit methods described in Chapter 7 and the ecoselection procedures of Chapters 8 and 9.

Aggregain (2008). Aggregain, developed and distributed by WRAP, is a free analysis tool running in Microsoft Excel for promoting the supply and use of recycled and secondary aggregates (including recycled concrete from construction, demolition waste material, and railway ballast) for the construction and road-building industries.

KCL-ECO 3.0. KCL represents the paper-making industry. KCL-Eco is an LCA tool designed specifically for this industry.

Eio-lca (2008). Economic input/output LCA (Eio-lca) of Carnegie Mellon University calculates sector emissions based on input/output data for the sectors of the North American Industry Classification Scheme (NAICS). It is not designed for the assessment of products. Demo available.

3.9 Exercises

E.3.1. Which phase of life would you expect to be the most energy intensive (in the sense of consuming fossil fuel) for the following products? Pick one and list the resources and emissions you think would be associated with each phase of its life along the lines of Figure 3.3.

- A toaster
- A two-car garage
- A bicycle
- A motorbike
- A refrigerator
- A coffeemaker
- An LPG-fired patio heater

E.3.2. Identify an appropriate functional unit for each of the products listed below. Think of the basic need the product provides—it is this that determines use—and list what you would choose, thinking of all from an environmental standpoint.

Product
- Washing machines
- Refrigerators
- Home heating systems
- Air conditioners
- Lighting
- Home coffeemaker
- Public transport
- Handheld hair dryers

E.3.3. What is meant by "externalized" costs and costs the are "internalized" in an environmental context? Now a moment of introspection: list three internalized costs associated with your lifestyle. Now list three

	Material	Manufacture	Transport	Use	Disposal
Material resources (*high use = 0, none = 4*)					
Energy use (*high use = 0, none = 4*)					
Global warming (*much CO_2 = 0, no CO_2 = 4*)					
Human health (*Toxic emissions or waste?*)					
Column totals					

Sum across column totals

FIGURE 3.12 *Streamlined LCA matrix.*

that are externalized. If your life is so pure that you have fewer than three, list some for the other people you know.

E.3.4. What, in the context of life-cycle assessment, is meant by "system boundaries"? How are they set?

E.3.5. Describe briefly the steps prescribed by the ISO 14040 Standard to guide life-cycle assessment of products.

E.3.6. What are the difficulties with a full LCA? Why would a simpler, if approximate, technique be helpful?

E.3.7. Pick two of the products listed in Exercise E.3.1 and, using your judgment, attempt to fill out the simplified streamlined LCA matrix below to give an environmentally responsible product rating. Make your own assumptions (and report them) about where the product was made and thus how far it has to be transported, and whether it will be recycled. Assign an integer between 0 (highest impact) and 4 (least impact) to each box and then sum to give an environmental rating, providing a comparison. Try the protocol.

■ *Material.* Is it energy-intensive? Does it create excessive emissions? Is it difficult or impossible to recycle? Is the material toxic? If the answer to these questions is *yes*, score 4. If the reverse, score 0. Use the intermediate integers for other combinations.

- *Manufacture.* Is the process one that uses much energy? Is it wasteful (meaning cut-offs and rejects are high)? Does it produce toxic or hazardous waste? Does make use of volatile organic solvents? If yes, score 4. If no, score 0, etc.

- *Transport.* Is the product manufactured far from its ultimate market? Is it shipped by air freight? If both yes, score 4. If no, score 0.

- *Use.* Does the product use energy during its life? Is the energy derived from fossil fuels? Are any emissions toxic? Is it possible to provide the use-function in a less energy-intensive way? Scoring as above.

- *Disposal.* Will the product be ent to land-fill at end of life? Does disposal involve toxic or long-lived residues? Scoring as above.

What difficulties did you have? Do you feel confident that the results are meaningful?

End of first life: a problem or a resource?

4.1 Introduction and synopsis

When stuff is useful, we show it respect and call it *material*. When the same stuff ceases to be useful, we lose respect for it and call it *waste*. Waste is deplorable, and it is much deplored, that from packaging particularly so. Is it inevitable? The short answer is *yes*; it is a consequence of one of the inescapable laws of physics—that entropy can only increase. A fuller answer is *yes*, but. The *but* has a number of aspects. That is what this chapter is about.

First, a calibration. We (the global *we*) are consuming materials at an ever-faster rate (Chapter 2). The first owner of a product, at end of life, rejects it as "waste." So waste, too, is generated at an ever-growing rate. What happens to it? In five words: *landfill, combustion, recycling, reengineering,* or *reuse.* That sounds comprehensive; it must be feasible to find a home for cast-off products in one of these? Ah, but. The capacity of a channel for dealing with products at end of first life must, to be effective,

Is this waste or is it a resource? (Image courtesy Envirowise—Sustainable Practices, Sustainable Profits, a U.K. government program managed by AEA Technology Plc.)

match the rate of rejection. Only one of the five has any real hope of achieving this goal. And then there is the economics. End of life is not simple.

To start at the beginning: why do we throw things away?

4.2 What determines product life?

The rapid turnover of products we see today is a comparatively recent phenomenon. In earlier times, furniture was bought with the idea that it would fill the needs not just of one generation but of several—treatment that, today, is reserved for works of art. A wristwatch, a gold pen—once these were things you used for a lifetime and then passed on to your children. No more. Behind all this is the question of whether the *value* of a product increases or decreases with age.

A product reaches the end of its life when it's no longer valued. The cause of death is, frequently, not the obvious one—that the product just stopped working. The life expectancy is the least of the following:[1]

- The *physical life,* meaning the time in which the product breaks down beyond economic repair

- The *functional life,* meaning the time when the need for the product ceases to exist

- The *technical life,* meaning the time at which advances in technology have made the product unacceptably obsolete

- The *economical life,* meaning the time at which advances in design and technology offer the same functionality at significantly lower operating cost

- The *legal life*—the time at which new standards, directives, legislation, or restrictions make the use of the product illegal

- Finally, the *loss of desirability*—the time at which changes in taste, fashion, or aesthetic preference render the product unattractive.

One obvious way to reduce resource consumption is to extend product life, making it more durable. But durability has more than one meaning; we've just listed six. Materials play a role in them all—something that we return to later. Accept, for the moment, that a product *has* reached the end of its life. What are the options?

---

[1] This list is a slightly extended version of one presented by D. G. Woodward, Life-cycle costing, Int. J. Project Management, 1997, Vol. 15, 335–344.

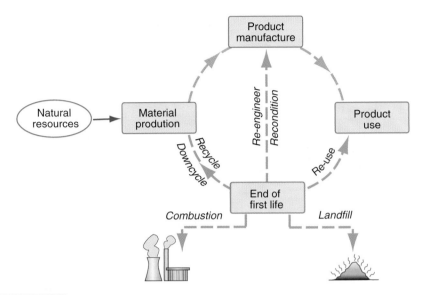

FIGURE 4.1 *End-of-life options: landfill, combustion, recycling, refurbishment or upgrading, and reuse.*

4.3 End-of-first-life options

Figure 4.1 introduces the options: landfill, combustion for heat recovery, recycling, reengineering, and reuse.

Landfill. Many of the products we now reject are committed to landfill. Already there is a problem; the land available to "fill" in this way is already, in some European countries, almost full. Recall one of the results discussed in Chapter 2: if the consumption of materials grows by 3% per year, we will use and, if we discard it, throw away as much "stuff" in the next 25 years as in the entire history of industrialization. Landfill is not going to absorb that. Governmental administrations react by charging a landfill tax—currently somewhere near €50 per tonne and rising, seeking to divert waste into the other channels of Figure 4.1. These must be capable of absorbing the increase. None, at present, can.

Combustion for heat recovery. Materials, we know, contain energy. Rather than throwing them away, it would seem better to retrieve and reuse some of this energy by controlled combustion, capturing the heat. But this is not as easy as it sounds. First there is the need for a primary sorting to separate combustible from noncombustible material (see Figure 4.2). Then the combustion must be carried out under conditions that do not generate toxic fumes or residues, requiring high temperatures, sophisticated control, and expensive equipment. The energy recovery is imperfect partly because it is

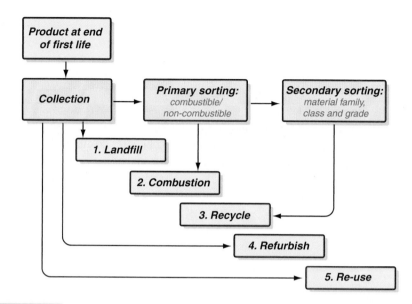

FIGURE 4.2 *End-of-life scenarios: landfill, combustion for heat recovery, recycling, refurbishment, and reuse. Different levels of sorting and cleaning are required for each.*

incomplete and partly because the incoming waste carries a moisture content that has to be boiled off. The efficiency of heat recovery from the combustion process is at best 50%, and if the recovered heat is used to generate electricity, it falls to 35%. And communities don't like an incinerator at their back door. Thus useful energy *can* be recovered by the combustion of waste, but the efficiency is low, the economics are unattractive, and the neighbors can be difficult.

Despite all this, combustion for heat recovery is in some circumstances practical and attractive. The most striking example is the cement industry, one that has an enormous energy budget (and CO_2 burden) because of the inescapable step of calcining in its production. Increasingly, combustion of vehicle tires and industrial and agricultural wastes are used as a heat source, reducing the demand on primary fuels, but not, of course, the attendant CO_2.

Recycling. Waste is only waste if nothing can be done to make it useful. It can also be a resource. Recycling is the reprocessing of recovered materials at the end of product life, returning them into the use stream. It is the end-of-life scenario that is best adapted to extracting value from the waste stream. We return to this topic in Section 4.5 for a closer look.

Reengineering or reconditioning. There is the story of the axe—an excellent axe—that, over time, had two new heads and three new handles. But it was still the same axe. Refurbishment, for some products, is cost effective and, compared with total replacement, energy efficient. Aircraft, for instance, don't wear out; instead, replacement of critical parts at regular inspection periods keeps a plane, like the axe, functioning just as it did when it was new. The Douglas DC-3, a 60-year-old design, is still flying, though not of course in the hands of its original owner. Premium airlines fly premium aircraft, so older models are sold on to operators with smaller budgets.

Reengineering is the refurbishment or upgrading of the product or of recoverable components. Certain criteria must be met to make it practical. One is that the design of the product is fixed, as it is with aircraft once an airworthiness certificate is issued, or that the technology on which it is based is evolving so slowly that there remains a market for the restored product. Some examples are housing, office space, and road and rail infrastructure; these are sectors with enormous appetites for materials. Some more examples: office equipment, particularly printing equipment and copying machines, and communication systems. These are services; the product providing them is unimportant to those who need the service, so long as it works well. It makes more sense to lease a service (as we all do with telephone lines, mobile phones, Internet service provision, and much else) because it is in the leasers' interests to maximize the life of the equipment.

And that is the other obstacle (together with rapidly changing technology) to refurbishment: that fashion, style, and perceptions change, making a refurbished product unacceptable, even though it works perfectly well. Personal image, satisfaction—even self-respect are powerful drivers.

Reuse. The cathedrals of Europe, almost all of them, are built on the foundations of earlier structures, often from the 10th or 11th century—built, in turn, on a still earlier 5th- or 6th-century beginning. If in a region that was once part of the Roman Empire, columns, friezes, fragments of the forum, and other structural elements of yet greater antiquity find their way into the structure, too. Reuse is not a new idea, but it is a good one.

Put more formally: reuse is the redistribution of the product to a consumer sector that is willing to accept it in its used state, perhaps to use for its original purpose (a secondhand car, for instance), perhaps to adapt to another (converting the car to a hot-rod or a bus into a mobile home).[2] That is a question of communication. Housing estate listings and used car and boat magazines exist precisely to provide channels of communication. Charity shops acquire clothing, objects, and junk[3] from those for whom

[2] What most of us see as waste can become the material of invention to the artist. For remarkable examples of this idea, visit the Museé International des Arts Modestes, 23 quai du Maréchal de Lattre de Tassigny, 34200 Sète, France (www.miam.org).

[3] I remember a sign above a store in an English town: "we buy junk. We sell antiques."

they had become waste, and sell them on to others who perceive them to have value. And here's a thought: the most effective tool ever devised to promote reuse is probably eBay.com, successful in this task precisely because it provides a global channel of communication.

4.4 The problem of packaging

Few applications of materials attract as much criticism as their use in packaging. Packaging ends its functional life as soon as the package is opened. It is ephemeral, it is trite, it generates mountains of waste, and most of the time it is unnecessary. Or is it? Think for a moment about the most highly developed form that packaging takes: the way we package ourselves. Clothes provide protection from heat and cold, from sun and rain. Clothes convey information about gender and ethnic and religious background. Uniforms identify membership and status, most obviously in the military and the church, but also in other hierarchical organizations: airlines, hotels, department stores, even utility companies. And at a personal level, clothes do much more: they are an essential part of the way we present ourselves. Though some people make the same clothes last for years, others wear them only once or twice before—for them—they become "waste" and are given to a charity or consignment shop.

Fine, you might say—we need packaging of that sort, but products are inanimate. What's the point of packaging for them? The brief answer: products are packaged for precisely the same reasons that we need clothes—protection, information, affiliation, status, and presentation.

So let us start with some facts. Packaging makes up about 18% of household waste but only 3% of landfill. Its carbon footprint is 0.2% of the global total. Roughly 60% of packaging in Europe, rather less in the United States, is recovered and used for energy recovery or recycling. Packaging makes possible the lifestyle we now enjoy. Without it, supermarkets would not exist. By protecting foodstuffs and controlling the atmosphere that surrounds them, packaging extends product life, allows access to fresh products all year round, and reduces food waste in the supply chain to about 3%; without packaging the waste is far higher. Tamperproof packaging protects the consumer. Pack information identifies the product and its sell-by date (if it has one) and gives instructions for use. Brands are defined by their packaging—the Coca-Cola bottle, the Campbell's soup can, Kellogg products—essential for product presentation and recognition.

The packaging industry[4] is well aware of the negative image that packaging, because of its visibility, holds, and it strives to minimize its weight and volume. There are, of course, exceptions, but there has been progress in optimizing it—providing all its functionalities with the minimum use of materials. The most used of these—paper, cardboard, glass, aluminium, and steel—have established recycling markets (see Section 4.5). Much packaging ends up in household waste, the most difficult to sort. The answer is better waste-stream management, in which the sorting is done by the consumer via marked containers. The protective function of much packaging requires material multilayers that cannot be recycled but that, if sorted, are still a source of energy.

Legislating packaging out of existence would require major adjustment of lifestyle, greatly increase the waste stream, and deprive consumers of convenience, product protection and hygienic handling. The challenge is that of returning as much of it as possible into the materials economy.

The Role of industrial design. What have you discarded lately that still worked or, if it didn't, could have been fixed? Changing trends, urged on by seductive advertising, reinforce the desire for the new and urge the replacement of still-useful objects. Industrial design carries a heavy responsibility here; it has, at certain periods, been directed toward creative obsolesce: designing products that are desirable only if new and urging the consumer to buy the latest models, using marketing techniques that imply that acquiring them is a social and psychological necessity.

But that is only half the picture. A well-designed product can acquire a value with age, and—far from becoming unwanted—can outlive its design life many times over. The auction houses and antique dealers of New York, London, and Paris thrive on the sale of products that, often, were designed for practical purposes but are now valued more highly for their aesthetics, associations, and perceived qualities. People do not throw away products for which they feel emotional attachment. So there you have it: industrial design both as villain and as hero. Where can it provide a lead?

When your house no longer suits you, you have two choices: you can buy a new house or you can adapt the one you have, and in adapting it you make it more personally yours. Houses allow this behavior. Most other products do not; and an old product (unlike an old house) is often perceived to be incapable of change and to have such low value that it is simply discarded.

---

[4] See, for instance, The Packaging Federation, www.packagingfedn.co.uk, or the Flexible Packaging Association, www.flexpack.org.

That highlights a design challenge: to create products that can be adapted and personalized so that they acquire, like a house, a character of their own and transmit the message, "Keep me, I'm part of your life." This suggests a union of technical and industrial design to create produces that can accommodate evolving technology but at the same time are made with a quality of material, design, and adaptability that creates lasting and individual character, something to pass on to your children.[5]

4.5 Recycling: resurrecting materials

Of the five end-of-life options shown in Figure 4.1, only one meets the essential criteria that:

- It can return waste materials into the supply chain

- It can do so at a rate that, potentially, is comparable with that at which the waste is generated

Landfill and combustion fail to meet the first, and refurbishment and reuse, almost always, fail the second. That leaves *recycling* (see Figure 4.2).

Quantification of the process of material recycling is difficult. Recycling costs energy, and this energy carries its burden of gases. But the *recycle energy* is generally small compared to the initial embodied energy, making recycling—when it is possible at all—an energy-efficient proposition. It might not, however, be one that is cost efficient; that depends on the degree to which the material has become dispersed. In-house scrap, generated at the point of production or manufacture, is localized and is already recycled efficiently (near 100% recovery). Widely distributed "scrap"—material contained in discarded products—is a much more expensive proposition to collect, separate, and clean. Many materials cannot be recycled, although they may still be reused in a lower-grade activity; continuous-fiber composites, for instance, cannot be reseparated economically into fiber and polymer to reuse them, though they can be chopped and used as fillers. Most other materials require an input of virgin material to avoid buildup of uncontrollable impurities. Thus the fraction of a material production that can ultimately reenter the cycle of Figure 4.1 depends on both the material itself and the product into which it has been incorporated.

Metals. The recycling of metals in the waste stream is highly developed. Metals differ greatly in their density, in their magnetic and electrical

---

[5] For an organization with such an ideal, see www.eternally-yours.nl.

properties, and even in their color, making separation comparatively easy. The value of metals, per kilogram, is greater than that of other materials. All these factors help make metal recycling economically attractive. There are many limitations on how recycled metals are used, but there are enough good uses that the contribution of recycling to today's consumption is large.

Polymers. The same cannot be said of polymers. Commodity polymers are used in large quantities, many in products with short life, and they present major problems in waste management, all of which, you would think, would encourage effective recycling. But polymers all have nearly the same density, have no significant magnetic or electrical signature, and can take on any color that the manufacturer likes to give them. They can be identified by X-ray fluorescence or infrared spectroscopy, but these methods are not infallible and they are expensive. Many are blends and contain fillers or fibers. The recycling process itself involves a large number of energy-consuming steps. Unavoidable contamination can prevent the use of recycled polymers in the product from which they were derived, restricting them to more limited use. For these reasons, the value of recycled polymers—the price at which they can be sold—is typically about 60% of that of virgin material.

A consequence of this situation is that the recycling of many commodity polymers is low. Increasing the recycle fraction is a question of identification, and here there is progress. Figure 4.3 shows, in the top row, the standard recycle marks, ineffective because they do not tell the whole story. The lower row shows the emerging identification system. Here polymer, filler, and weight fraction are all identified. The string, built up from the abbreviations listed in the appendix to this chapter, give enough information for effective recycling.

The economics of recycling. Although recycling has far-reaching environmental and social benefits, it is market forces that—until recently—determined whether or not recycling happened. Municipalities collect recyclable waste, selling it through brokers to secondary processors who reprocess the materials and sell them, at a profit, to manufacturers. The recycling market is like any other, with prices that fluctuate according to the balance of supply and demand. In a free market, the materials that are recycled are those from which a profit can be made. These include almost all metals but few polymers (see Table 4.1).

Scrap arises in more than one way. *New* or *primary scrap* is the cut-offs from billets, risers from castings, and turnings from machining that are a by-product of the manufacture of products; it can be recycled immediately, often in-house. *Old* or *secondary scrap* appears when the products

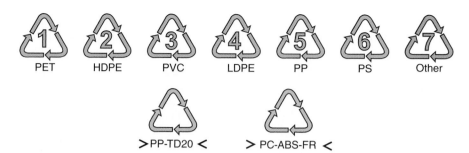

FIGURE 4.3 *Top: recycle marks for the most commonly used commodity polymers. Bottom: more explicit recycle marks detailing blending, fillers and reinforcement. The first is polypropylene 20% talc powder. The second is a polycarbonate/ABS blend with glass fiber. The coding is explained in the appendix to this chapter.*

Table 4.1 Recycling markets

Material family	Developed end uses for recycled materials	Existing secondary uses but not developed as a market
Metals	Steel and cast iron Aluminum Copper Lead Titanium All precious metals	Paper—metal foil packaging
Polymers and elastomers	Polyethyleneterephthalate (PET) High-density polyethylene (HDPE) Polypropylene (PP) Polyvinylchloride (PVC)	All other polymers and elastomers, notably tires
Ceramics and glasses	Bottle glass Brick Concrete and asphalt	Nonbottle glass
Other materials	Cardboard, paper, newsprint	Wood Textiles: cotton, wool, and other fibers

themselves reach the end of their useful lives (see Figure 4.4). The value of recyclable waste depends on its origin. New scrap carries the highest value because it is uncontaminated and easy to collect and reprocess. Old scrap from commercial sources such as offices and restaurants is more valuable than that from households because it is more homogeneous and needs less sorting.

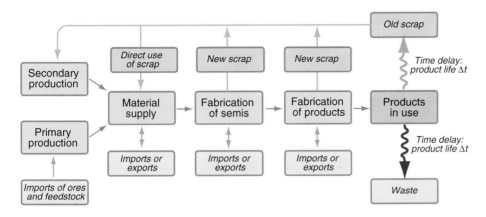

The material flows in the economy, showing recycling paths. New scrap arises during manufacture and is reprocessed almost immediately. Old scrap derives from products at end of life; it reenters production only after a delay of Δt, the product life.

Producers of secondary materials must, of course, compete with those producing virgin materials. It is this fact that couples the price of the first to that of the second. Virgin materials are more expensive than those that have been recycled, because their quality, both in engineering terms and that of perception, is greater. Manufacturers using recycled materials require assurance that this drop in quality will not compromise their products.

The profitability of a market can be changed by economic intervention—subsidies, for instance, or legislation with penalties for failure to comply. Legislation setting a required level of recycling of vehicles and of electronic products at the end of their lives is now in force in Europe; other nations have similar programs and plans for more. Municipalities, too, have recycling laws requiring the reprocessing of waste that, under free-market conditions, would have zero value. In a free market these products would end up in landfill, but the law prohibits it. When this is so, municipalities sell the materials for a negative price—that is, they pay processing firms to take them. The negative price, too, fluctuates according to market forces and may, if technology improves or demand increases, turn positive, removing the need for the subsidy.

Where laws requiring recycling do not exist, recycling must compete also with landfill. Landfill, too, carries a cost. What is recycled and what is dumped then changes as market conditions—the level of a landfill tax, for instance—change and businesses seek to minimize the cost of managing waste.

All this could give the impression that waste management is a local issue, driven by local or national market forces. But the insatiable appetite

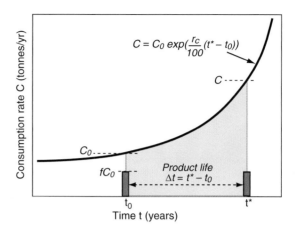

$$C = C_0 \exp(\frac{r_c}{100}(t^* - t_0))$$

FIGURE 4.5   *Material dispersed as products does not appear as scrap for recycling until the product comes to the end of life. If consumption grows, long-lived products contribute less than those with short lives.*

of the fast-developing nations, particularly China and India, turn the "waste" of Europe and the United States into what, for these developing countries, is a resource. Low labor costs, sometimes less restrictive environmental regulation, and different manufacturing quality standards drive a world market in both waste and recycled materials.

The contribution of recycling to current supply.   Suppose that a fraction f of the material of a product with a life of Δt becomes available as old scrap. Its contribution to today's supply is the fraction f of the consumption Δt years ago. Material consumption, generally, grows with time, so this delay between consumption and availability as scrap reduces the contribution it makes to the supply of today.

Figure 4.5 illustrates this idea. Suppose, for the moment, that a material exists that is used for one purpose only in a product with a life span Δt and that, at end of life, a fraction f (about 0.6 in the Figure) is recycled into supply for current consumption, which has been growing at a rate r_c % per year. If the consumption rate when the product was made at time t_0 was C_0 tonnes per year, the consumption today, at time t^*, is:

$$C = C_0 \exp(\frac{r_c}{100}(t^* - t_0)) = C_0 \exp(\frac{r_c}{100} \Delta t) \tag{4.1}$$

where $\Delta t = t^* - t_0$. The recovered fraction is:

$$R = f C_0 \tag{4.2}$$

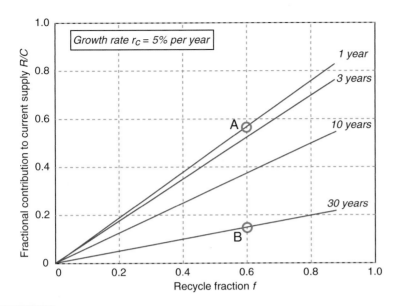

Plot of Equation 4.3 for recycling effectiveness when the growth rate in consumption is 5% per year, for various product lives and recycle fractions.

shown as a red bar on the Figure. Its fractional contribution to supply today is:

$$\frac{R}{C} = \frac{f}{exp(\frac{r_c}{100}\Delta t)} \qquad (4.3)$$

Figure 4.6 shows what this looks like for a product with a growth rate of 5% per year. If the fraction $f = 0.6$ and the product life Δt is one year, the contribution is large, about 0.58 of current supply (Point A on Figure 4.6). But if the product life is 30 years, the contribution falls to 0.17 (Point B). The recycle contribution increases with f, of course. But it decreases quickly if the product has a long life or a fast growth rate.

In reality most materials are used in many products, each with its own life span Δt_i, recycle fraction f_i, and growth rate $r_{c,i}$. Consider one of these— product "i"—that accounts for a fraction s_i of the total consumption of the material. Its fractional contribution to consumption today is:

$$\frac{R_i}{C} = \frac{s_i f_i}{exp(\frac{r_{c,i}}{100}\Delta t_i)} \qquad (4.4)$$

The total contribution of recycling is the sum of terms like this for the material in all the products that use it.

An example will bring out some of the features. A material is used to make both gizmos and widgets; each accounts for 25% of the total consumption of the material. Gizmos last for 20 years; their sales have grown steadily at 10% per year. At the end of life, gizmos are dismantled and all the material is recovered (f_{gizmo} = 1). Widgets, on the other hand, have an average life of four weeks and are difficult to collect; their recycle fraction is only f_{widget} = 0.5. Their sales have grown slowly, at 1% per year for the recent past. Inserting these data into Equation 4.4, we find an unexpected result: Gizmos, all of which are recycled, contribute the tiny fraction of 0.03 to current supply, whereas widgets, only half of which are recycled, contribute a much larger fraction of 0.125. The main effect here is the product lifetime: products with short lives make larger contributions to recycling than those that last for a long time. This reveals one of the many unexpected aspects of the materials economy: making products that last longer can reduce demand, but it also reduces the scrap available for recycling. If, in making products last longer, they have to be made more robust, meaning that they use more material, the effect can even be to increase the demand for primary materials.

4.6 Summary and conclusion

The greater the number of us that consume and the greater the rate at which we do so, the greater is the volume of materials that our industrial system ingests and then ejects as waste. Real waste is a problem—a loss of resources that cannot be replaced, and there has to be somewhere to put it and that, too, is a diminishing resource.

But waste can be seen differently: as a resource. It contains energy and it contains materials, and—since most products still work when they reach the end of their first life—it contains components or products that are still useful. There are a number of options for treating a product at the end of its first life: extract the energy via combustion, extract the materials and reprocess them, replace the bits that are worn, and sell it again, or, more simply, put it on eBay or another trading system and sell it as is.

All these options have merit. But only one—recycling—can begin to cope with the volume of waste that we generate and transform it into a useful resource.

4.7 Further reading

Chapman, P.F. and Roberts, F. (1983), "Metal resources and energy, Butterworth's Monographs in Materials", Butterworth and Co., ISBN 0-408-10801-0. *(A monograph that analyzes resource issues, with particular focus on energy and metals.)*

Chen, R.W., Navin-Chandra, D. Prinz, F.B. (1993), "Product design for recyclability: a cost/benefit analysis". Proceedings of the IEEE International Symposium on Electronics and the Environment, Vol. 10–12, 178–183, ISEE.1993.302813. *(Recycling lends itself to mathematical modeling. Examples can be found in the book by Chapman and Roberts, above, and in this paper, which takes a cost-benefit approach.)*

Guidice, F., La Rosa, G. and Risitano, A. (2006), "Product design for the environment", CRC/Taylor and Francis, ISBN 0-8493-2722-9. *(A well-balanced review of current thinking on eco-design.)*

Henstock, M.E. (1998), "Design for recyclability", Institute of Metals. *(A useful source of background reading on recycling.)*

4.8 Appendix: designations used in recycle marks

Designations used in recycle marks: base polymers

E/P	ethylene-propylene plastic
EVAC	ethylene-vinyl acetate plastic
MBS	methacrylate-butadiene-styrene plastic
ABS	acrylonitrile-butadiene-styrene plastic
ASA	acrylonitrile-styrene-acrylate plastic
C	cellulose polymers
COC	cycloolefin copolymer
EP	epoxide; epoxy resin or plastic
Imod	Impact modifier
LCP	liquid-crystal polymer
MABS	methacrylate-acrylonitrile-butadiene-styrene plastic
MF	melamine-formaldehyde resin
MPF	melamine-phenolic resin
PA11	Homopolyamide (Nylon) based on 11-aminoundecanoic acid
PA12	homopolyamide (Nylon) based on ω-aminododecanoic acid or on laurolactam
PA12/MACMI	copolyamide(Nylon) based on PA12, 3, 3-Dimethyl-4, 4-diaminodicyclo-hexylmethane and isophthalic acid
PA46	homopolyamide (Nylon) based on tertramethylenediamine and adipic acid
PA6	homopolyamide (Nylon) based on ε-caprolactam
(Ion)0	homopolyamide (Nylon) based on hexamethylenediamine and sebacic acid
PA612	homopolyamide (Nylon) based on hexamethylenediamine and dodecane-diacid (1,10-Decandicarboxylic acid)
PA66	homopolyamide (Nylon) based on hexamethylenediamine and adipic acid

PA66/6T	copolyamide based on hexamethylenediamine, adipic acid and terephthalic acid
PA666	copolyamide based on hexamethylenediamine, adipic acid and ε-caprolactam
PA6I/6T	copolyamide based on isophthalic acid, adipic acid and terephthalic acid
PA6T/66	copoylamide based on adipic acid, terephthalic acid and hexamethylenediamine
PA6T/6I	copolyamide based on hexamethylenediamine, terephthalic acid, adipic acid and isophthalic acid
PA6T/XT	copolyamide based on hexamethylenediamine, 2-methyl-penta-mehylene diamine and terephthalic acid
PAEK	polyaryletherketon
PAIND/INDT	copolyamide based on 1, 6-diamino-2, 2, 4-trimethylhexane, 1, 6-diamino-2, 4, 4-trimethylhexane and terephthalic acid
PAMACM12	homopolyamide based on 3,3'-dimethyl-4,4'-diaminodicyclohexyl-methane and dodecandioic acid
PAMXD6	homopolyamide based on m-xylylenediamine and adipic acid
PBT	poly(butylene terephthalate)
PC	polycarbonate
PCCE	poly(cyclohexylene dimethylene terephthalate
PCTA	poly(cyclohexylene dimethylene terephthalate, acid
PCTG	poly(cyclohexylene dimethylene terephthalate, glycol
PE	polyethylene
PEI	polyetherimide
PEN	polyethylene naphthalate
PES	polyethersulfone
PET	polyethylene terephthalate
PETG	polyethylene terephthalate, glycol
PF	phenol-formaldehyde resin
PI	polyimide
PK	polyketone
PMMA	poly(methyl methacrylate)
PMMI	Polymethylmethacrylimide
POM	polyoxymethylene, polyacetale, polyformaldehyde
PP	polypropylene
PPE	poly(phenylene ether)
PPS	poly(phenylene sulfide)
PPSU	poly(phenylene sulfone)
PS	polystyrene
PS-SY	polystyrene, syndiotactic

PSU	polysulfone
PTFE	polytetrafluoroethylene
PUR	polyurethane
PVC	polyvinyl chloride
PVDF	poly(vinylidene fluoride
SAN	styrene-acrylonitrile plastic
SB	styrene-butadiene plastic
SMAH	styrene-maleic anhydride plastic
TEEE	thermoplastic ester- and ether-elastomers
TPA	polyamide thermoplastic elastomer
TPC	copolyester thermoplastic elastomer
TPO	olefinic thermoplastic elastomer
TPS	styrenic thermoplastic elastomer
TPU	urethane thermoplastic elastomer
TPV	thermoplastic rubber vulcanisate
TPZ	unclassified thermoplastic elastomer
UP	unsaturated polyester

Designations used in recycle marks: fillers

CF	carbon fiber
CD	carbon fines, powder
GF	glass fiber
GB	glass beads, spheres, balls
GD	glass fines, powder
GX	glass not specified
K	calcium carbonate
MeF	metal fiber
MeD	metal fines, powder
MiF	mineral fiber
MiD	mineral fines, powder
NF	natural organic fiber
P	mica
Q	silica
RF	aramid fiber
T	talcum
X	not specified
Z	others not included in this list

4.9 Exercises

E.4.1. Many products are thrown away and enter the waste stream, even though they still work. What are the reasons for this?

E.4.2. Do you think manufacture without waste is possible? "Waste", here, includes waste heat, emissions, and solid and liquid residues that cannot be put to a useful purpose. If not, why not?

E.4.3. What options are available for coping with the waste stream generated by modern industrial society?

E.4.4. Recycling has the attraction of returning materials into the use stream. What are the obstacles to recycling?

E.4.5. Car tires create a major waste problem. Use the Internet to research ways in which the materials contained in car tires can be used, either in the form of the tire or in some decomposition of it.

E.4.6. List three important functions of packaging.

E.4.7. As a member of a brainstorming group, you are asked to devise ways of reusing polystyrene foam packaging—the sort that encases TV sets, computers, appliances, and much else when transported. Use free thinking; no suggestion is too ridiculous.

E.4.8. You are employed to recycle German washing machines, separating the materials for recycling. You encounter components with the following recycle marks:

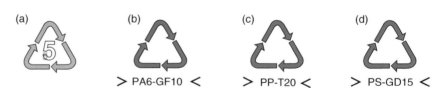

(a)   (b) > PA6-GF10 <   (c) > PP-T20 <   (d) > PS-GD15 <

How do you interpret them?

Answer. (a) Polypropylene. (b) Polyamide 6 (Nylon 6) with 10% glass fiber. (c) Polypropylene with 20% talc. (d) Polystyrene with 15% glass fines (powdered glass).

E.4.9. The metal lead has a number of uses, principally as electrodes in vehicle batteries, in architecture for roofing and pipe work (particularly on churches), and as pigment for paints. The first two of these allow recycling; the third does not. Batteries consume 38% of all lead, have an average life of four years, and have a growth rate of 4% per year, and the lead they contain is recycled with an efficiency of 80%. Architectural lead accounts for 16% of total consumption. The lead on buildings has an average life of 70 years, the same growth rate (4% per year) as batteries, and 95% of it is recycled. What is the fractional contribution of recycled lead from each source to current supply?

E.4.10. A material M is imported into a country principally to manufacture one family of products with an average life span of five years and a growth rate of r_c % per year. The material is not at present recycled at end of life, but it could be. The government is concerned that imports should not grow. What level of recycling is necessary to make this possible?

CHAPTER 5

The long reach of legislation

CONTENTS

5.1 Introduction and synopsis

The prophet Moses, seeking to set standards for the ways in which his people behaved, created or received (according to your viewpoint) 10 admirably concise commandments. Most start with the words "Thou shalt not …," with simple, easily understood incentives (heaven, hell) to comply. Today, as far as materials and design are concerned, it is Environmental Protection Agencies and European Commissions that issue commandments, or, in their language, *directives*. The consequences of infringing them are not as Old Testament in their severity as those of the original 10, but if you want to grow your business, compliance becomes a priority.

This involves some obvious steps:

■ Being aware of directives or other binding controls that touch on the materials or processes you use

---

Warning signs that relate to materials. Clockwise from top left: dangerous, highly flammable, explosive, poisonous, environmentally hazardous, very corrosive.

- Understanding what is required to comply with them

- Having (or developing) tools to make compliance as painless as possible

- Exploring ways to make compliance profitable rather than a burden; exploiting compliance information as a marketing tool, for example

This chapter is about controls and economic instruments that impinge on the use of engineering materials. It reviews current legislation and describes an example of tools to help with compliance.

5.2 Growing awareness and legislative response

Table 5.1 lists nine documents that have had profound influence on current thinking about the effects of human activity on the environment. The publications span a little less than 50 years. Over this period the approach to pollution and environmental law has evolved through a number of phases,[1] best summarized in the following way:

- Ignore it: pretend it isn't there

- Dilute it: make the smokestack taller or pump it further out to sea

- Fix it where it is a problem: the "end-of-pipe" approach

- Prevent it in the first place: the first appearance of design for the environment

- Sustainable development: life in equilibrium with the environment—the phase we are in now

Current thinking has stimulated national legislation and international protocols and agreements. The international agreements tend to be broad statements of intent. The national legislation, by contrast, tends to be specific and detailed.

Historically, environmental legislation has targeted individual, obvious problems—dumping of toxic waste, lead in petrol, water pollution, ozone depletion—taking a *command and control* approach: "Thou shalt not" cast in modern terms. There is a growing recognition that this approach can lead to perverse effects, where action to fix one isolated problem simply

---

[1] Detailed in books on industrial ecology such as Ayres and Ayres (2002); see Further Reading.

Table 5.1	Required reading: landmark publications
Date, author, and title	**Subject**
1962: Rachel Carson, *Silent Spring*	Meticulous examination of the consequences of the use of the pesticide DDT and of the impact of technology on the environment.
1972: Club of Rome, *Limits to Growth*	The report that triggered the first of a sequence of debates in the 20th century on the ultimate limits imposed by resource depletion.
1972: The Earth Summit in Stockholm	The first conference convened by the United Nations to discuss the impact of technology on the environment.
1987: The UN World Commission on Environment and Development (WCED), *Our common future*	Known as the Brundtland Report, it defined the principle of sustainability as "Development that meets the needs of today without compromising the ability of future generations to meet their own needs."
1987: Montreal Protocol	The International Protocol to phase out the use of chemicals that deplete ozone in the stratosphere.
1992: Rio Declaration	An international statement of the principles of sustainability, building on those of the 1972 Stockholm Earth Summit.
1998: Kyoto Protocol	An international treaty to reduce the emissions of gases that, through the greenhouse effect, cause climate change.
2001: Stockholm Convention	The first of ongoing meetings to agree on an agenda for the control and phase-out of persistent organic pollutants (POPs).
2007: IPCC Fourth Assessment Report, *Climate Change 2007: The Physical Basis*	This Report of the Intergovernmental Panel on Climate Change (IPCC) establishes beyond any reasonable doubt the correlation between carbon in the atmosphere and climate change.

shifts the burden elsewhere and may even increase it. For this reason there has been a shift from command and control legislation toward the use of *economic instruments*—green taxes, subsidies, trading schemes—that seek to use market forces to encourage the efficient use of materials and energy. We have already seen that some activities create environmental burdens that have costs that are not paid for by the provider or user. These are called external costs, or *externalities*. A more effective approach is to transfer the costs back to the activity creating it, thereby *internalizing* them.

5.3 International treaties, protocols, and conventions

It is exceedingly difficult to negotiate enforceable treaties that bind all the nations of the planet to a single course of action; the diversity of culture, national priorities, economic development, and wealth are too great. The best the international community can achieve is an Agreement, Declaration of Intent, or Protocol[2] that a subset of nations feels able to sign. Such agreements directly influence policy in the nations that sign them. And by defining the high ground they exert moral pressure on both signatories and nonsignatories alike. Two are particularly significant in their influence on government policy on materials:

The Montreal Protocol (1989) is a treaty aimed at reducing the use of substances that deplete the ozone layer of the stratosphere. Ozone depletion allows more UV radiation to reach the surface of the Earth, damaging living organisms. The culprits are typified by CFCs—chlorofluorocarbons—that were widely used as refrigerants and as blowing agents for polymer foams, particularly those used for house insulation. The Protocol has largely achieved its aims.

The Kyoto Protocol (1997) is an international treaty to reduce the emissions of gases that, through the greenhouse effect, cause climate change. It sets binding targets for 37 industrialized countries and the European community that have signed it, committing them to reduce greenhouse gas emissions over the five-year period 2008–2012.

International directives and protocols are usually based on principles—statements of what are seen as fundamental rights—rather than on laws that cannot be agreed or enforced. Here are examples of some that have emerged from the Protocols and Conventions listed in Table 5.1:

- *Principle 21 (Stockholm Declaration).* The right to exploit one's own environment.

- *Principle 2 (Rio Declaration).* The right to development without damage to others.

- *Precautionary principle (WCED report).* Where there are possibilities of large irreversible impacts, the lack of scientific certainty should not stop preventative action from being taken.

- *Polluter pays principle.* Transfers the responsibility and cost of pollution to the polluter.

---

[2] A protocol is defined as "a memorandum of resolutions arrived at in negotiation, signed by the negotiators, as a basis for a final convention or treaty." In fact, the Kyoto Protocol is more than that, being a binding treaty to meet certain agreed objectives. The distinction between the Protocol and the Convention, in current usage, is that though the Convention *encourages* countries to stabilize emissions, the Protocol *commits* them to do so.

■ *The principle of sustainable development.* Protection of the environment, equity of burden.

The idea is that these should provide a framework within which strategies and actions are developed.

5.4 National legislation: standards, directives, taxes, trading tools

"The Council of the European Union, having regard to A, B, and C, acting in accordance with procedures P, Q, and R of activities X, Y and Z, and whereas ... (there follows a list of 27 "whereases") *has adopted this directive* ..."—that was a paraphrase of the way an EU Directive starts. Environmental legislation makes heavy reading. These directives are cast in legal language of such Gothic formality and Baroque intricacy that organizations spring up with the sole purpose of interpreting it. But since much of it impinges, directly or indirectly, on the use of materials, it is important to get the central message.

National legislation, typified by U.S. Environmental Agency Acts or European Union Environmental Directives, takes four broad forms:

■ Setting up standards

■ Voluntary agreements negotiated with industry

■ Binding legislation that imposes requirements with penalties if they are not met

■ Economic instruments that seek to use market forces to induce change: taxes, subsidies, and trading schemes

Standards. ISO 14000 of the International Standards Organization defines a family of standards for environmental management systems.[3] It contains the set IS0 14040, 14041, 14042, and 14043, published between 1997 and 2000 and prescribing broad but vague procedures for the four steps described in Chapter 3, Section 3.3 (*goals and scope, inventory compilation, impact assessment,* and *interpretation*). The standard is an attempt to bring uniform practice and objectivity into life-cycle assessment and its interpretation, but it is not binding in any way.

---

[3] See www.iso-14001.org.uk/iso-14040 for a summary.

ISO 14025 is a standard guiding the reporting of LCA data as an Environmental Product Declaration (EPD) or a Climate Declaration (CD).[4] The goal is to communicate information about the environmental performance of products as a "declaration" in a standard, easily understood format. The data used for the declaration must follow the ISO 14040 family of procedures and must be independently validated by a third party. The EPD describes the output of a full LCA or (if declared so) part of one. The CD is limited to emissions that contribute to global warming: CO_2, CO, CH_4, and N_2O.

In practice, LCA procedures are used primarily for in-house product development, benchmarking, and promoting the environmental benefits of one product over another. They are rarely used as the basis for regulation because of the difficulties, described in Chapter 3, of setting system boundaries, of double counting, and of limited coverage across products.

Voluntary agreements and binding legislation. Current legislation is aimed at internalizing costs and conserving materials by increasing manufacturers' responsibilities, placing on them the burden of cost for disposal. Here are some examples.

The U.S. Resource Conservation and Recovery Act (RCRA), enacted in 1976, is a federal law of the United States. The Environmental Protection Agency (EPA) monitors compliance. RCRA's goals are to:

- Protect the public from harm caused by waste disposal

- Encourage reuse, reduction, and recycling

- Clean up spilled or improperly stored wastes

The U.S. EPA 35/50 Program (1988) identified 16 priority chemicals, italicized in Table 5.2, with the aim of reducing industrial toxicity by voluntary action by industry over a 10-year period.

Other legislation includes:

- *The U.S. EPA Code of Federal Regulation (CFR).* Protection of the environment (CFR Part 302) deals with protection of the environment and human health, imposing restrictions on chemicals released into the environment during manufacture, life, and disposal. Like REACH (discussed later in this list), it requires manufacturers to register the use of any one of a long list of chemicals and materials (Table 302.2 of the Regulation) if the quantity used exceeds a threshold.

---

[4]For the more intimate details, see www.environdec.com

Table 5.2	Priority chemicals and materials
Volatile organic compounds (VOCs)	**Applications**
Benzene	Intermediate in production of styrene, thus many polymers
Carbon tetrachloride	Solvent for metal degreasing, lacquers
Chloroform	Solvent
Methyl ethyl ketone	Solvent for metal degreasing, lacquers
Tetrachoroethylene	Solvent for metal degreasing
Toluene	Solvent
Trichlorethylene	Solvent, base of adhesives
Xylenes	Lacquers, rubber adhesives
Toxic metals or salts of metals	
Asbestos	Fibro-board reinforcement, thermal and electrical insulation
Antimony	Bearings, pigments in glasses
Beryllium + compounds	Space structures, copper-beryllium alloys
Cadmium and its compounds	Electrodes, plating, pigment in glasses and ceramics
Chromium compounds	Electroplating, pigments in glasses and ceramics
Cobalt + and compounds	Superalloys, pigments in glasses and glazes
Lead + compounds	Storage batteries, bearing alloys, solders
Mercury + compounds	Control equipment, liquid electrode in chemical production
Nickel + compounds	Nickel carbonyl as intermediate in nickel production
Radioactive materials	Materials science and medicine
Toxic chemicals	
Cyanides	Electroplating, extraction of gold and silver

- *Volatile Organic Compounds (VOCs, 1999).* The European Directive EC 1999/13 aimed to limit the emissions of VOCs caused by the use of organic solvents and carriers like those in organic-based paints and industrial cleaning fluids. Compliance became mandatory in 2007.

- *End-of-Life Vehicles (ELV, 2000).* The European Community Directive EC2000/53 establishes norms for recovering materials from dead cars. The initial target, a rate of reuse and recycling of 80% by weight of the vehicle and the safe disposal of hazardous materials, was established in 2006. By 2015 the recycling target rises to 85%. The motive is to encourage manufacturers to redesign their products to avoid using hazardous materials and to design them to maximize ease of recovery and reuse.

- *Hazardous Substances Directive (RoHS 2002).* The RoHS Directive stands for "the restriction of the use of certain hazardous substances in electrical and electronic equipment." This directive bans the placing on the EU market of new electrical and electronic equipment containing more than agreed levels of six materials: lead, cadmium, mercury, hexavalent chromium, polybrominated biphenyl (PBB), and polybrominated diphenyl ether (PBDE) flame retardants. It is closely linked with the WEEE Directive (discussed next), which sets collection, recycling, and recovery targets for electrical goods and is part of a legislative initiative to solve the problem of huge amounts of toxic waste arising from electronic products.

- *Waste Electrical and Electronic Equipment (WEEE,2002).* A similar directive (EC 2002/96 and 2003/108) on waste electrical and electronic equipment seeks to increase recovery, recycling, and reuse of electronic equipment and electrical appliances. It requires that producers finance the collection, recovery, and safe disposal of their products and meet certain recycling targets. Products failing to meet the requirement must be marked accordingly (a crossed-out wheeled bin).

- *Energy-using Products (EuP, 2003).* The EU Directive EC 2003/0172 establishes a framework of ecodesign requirements for products that use energy—appliances, electronic equipment, pumps, motors, and the like. It requires that manufacturers of any product that uses energy "shall demonstrate that they have considered the use of energy in their product as it relates to materials, manufacture, packaging transport and distribution, use, and end of life. For each of these the consumption of energy must be assessed and steps to minimise it identified."

- *Registration, Evaluation, Authorization, and Restriction of Chemical Substances (REACH, 2006).* The Directive EC 1907/2006 came into force in June 2007, to be phased in over the following 11 years. The directive places more responsibility on manufacturers to manage risks from chemicals and to find substitutes for those that are most dangerous. The list is a long one—it has some 30,000 chemicals on it—and it affects anyone in the European Union who produces, trades, processes, or consumes any "chemical," including metals and alloys, in quantities greater than 1 tonne per year. And there is more. Manufacturers in Europe and importers into Europe must register the restricted substances they use by providing a detailed technical dossier for each, listing their properties, an assessment of their impacts on the environment and human health, and the risk-reduction measures they have adopted. Without preregistration, it is illegal for manufacturers and importers to place substances on the market.

5.5 Economic instruments: taxes and trading schemes

"Economic instruments manipulate market forces to influence the behaviour of consumers and manufacturers in ways that are more subtle and effective than conventional controls, and they generally do so at lower cost."[5] Well, that's the idea. Taxation, one "economic instrument," might not strike you as subtle,[6] but it does seem to work better than "command and control" methods. Here are some examples.

Green taxes. Many countries now operate a *landfill tax*, currently standing at around €50 (US$80) per tonne in Europe, as a tool to reduce waste and foster better waste management. An *aggregate tax* on gravel and sand, about €2 (US$3) per tonne, recognizes that extracting aggregate has environmental costs; it is designed to reduce the use of virgin aggregate and stimulate the use of waste from construction and demolition. Most nations impose a *fuel duty*—a tax on gasoline and diesel fuel—to encourage a shift to fuel-efficient vehicles and increase the use of biofuels by taxing them at a lower rate. Increasingly governments impose a *carbon tax* (at present on the order of €100, or US$160, per tonne of carbon), often based on energy consumption (using energy as a proxy for carbon), and NO_x and SO_x *emission taxes*. Roughly half the U.S. states charge a deposit, a returnable tax, on bottles and cans, a scheme that has proved very effective in returning these materials into the recycling loop.

---

[5]U.K. Department for Environment, Food, and Rural Affairs (DEFRA), www.defra.gov.uk/environment/index.html.

[6]"Nothing is certain but death and taxes"—that was Benjamin Franklin writing 250 years ago.

But imposing a tax—a carbon tax, for instance—has two difficulties. First, it does not guarantee the environmental outcome of reducing CO_2; industries that can afford it will simply pay it. The second is one of public acceptance. Taxes carry high administration costs, and people don't trust governments to spend the tax on the environment; fuel taxes, for example, don't get spent on roads or pollution-free vehicles. Of those two certainties of life, death and taxes, it is taxes that people try hardest to avoid.

Trading schemes.   Another way of putting a value on something is to create a market for it. The stock market is an example: a company issues shares, the total number of which represents its "value." The shares are traded (sold or purchased for real money) and they therefore float in value, rising if they are seen as undervalued, falling if they are seen as overvalued. At any moment in time the share price sets a value on the company. A notion emerging from the Kyoto Meeting of 1987 was to adapt this "market principle" to establish a value for emissions. To see how it works and the difficulties with it, we need to digress to explore emissions trading.

Emissions trading is a market-based scheme that allows participants to buy and sell permits for emissions, or credits for reduction in emissions (a different thing) in certain pollutants (those of global impact, such as CO_2). Taking carbon as an example, the regulator first decides on a total acceptable emissions level and divides this into tradable units called *permits*. These are allocated to the participants, based on their actual carbon emissions at a chosen point in time. The actual carbon emissions of any one participant change with time, falling if they develop more efficient production technology or rising if they increase capacity. A company that emits more than its allocated allowances must purchase allowances from the market, whereas a company that emits less than its allocations can sell its surplus. Unlike regulation that imposes emission limits on particular facilities, emissions trading gives companies the flexibility to develop their own strategy to meet emission targets, by reducing emissions on site, for example, or by buying allowances from other companies that have excess allowances. The environmental outcome is not affected, because the total number of permits is fixed or is reduced over time as environmental concerns grow. The buyer is paying a charge for polluting while the seller is rewarded for having reduced emissions. Thus those who can easily reduce emissions most cheaply will do so, achieving pollution reduction at the lowest cost to society.

Emissions trading has another dimension—that of offsetting carbon release by buying *credits* in activities that absorb or sequester carbon or that replace the use of fossil fuels by energy sources that are carbon-free: tree planting, solar, wind, or wave power, for example. By purchasing sufficient credits, the generator of CO_2 can claim to be "carbon neutral." However, offsetting has its critics. Three of the more telling criticisms are:

- Offsetting provides an excuse for enterprises to continue to pollute as before by buying credits and passing the cost on to the consumer.

- The scheme only achieves its aim if the mitigating project runs for its planned life, and this is often very long. Trees, for instance, have to grow for 50 to 80 years to capture the carbon with which they are credited. Fell them sooner for quick profit and the offset has not been achieved. Wind turbines and wave power, similarly, achieve their claimed offset only at the end of their design life, typically 25 years.

- It is hard to verify that the credit payments actually reach the mitigating projects—the tree planters or wind turbine builders—for which they were sold; too much of it gets absorbed in administrative costs.

5.6 The consequences

The burden this legislation places on the materials and manufacturing industries is considerable. The requirements are far-reaching:

- Documentation of the use of any one of 30,000 listed chemicals

- Analysis of energy and material use in all energy-using products

- Finding substitutes for VOCs and other restricted substances

- Mandatory take-back, disassembly, and acceptable disposal of an increasingly large range of products

Figure 5.1 summarizes these interventions, suggesting where they influence the flows of the life cycle. The intention of the legislation—that of reducing resource consumption and damaging emissions and particularly of internalizing the costs these generate—makes sense. The difficulty is that implementing it generates administration, reporting, and other costs in addition to the direct costs of cleanup, thereby adding to the burden on industry.

These additional costs are minimized by the use of well-designed software and other tools. Most of the LCA packages documented in the appendix of Chapter 3 were created to help companies analyze their products, using standards that meet the requirements of the legislation. A Web search on any one of the acts and directives listed in Section 5.4 reveals

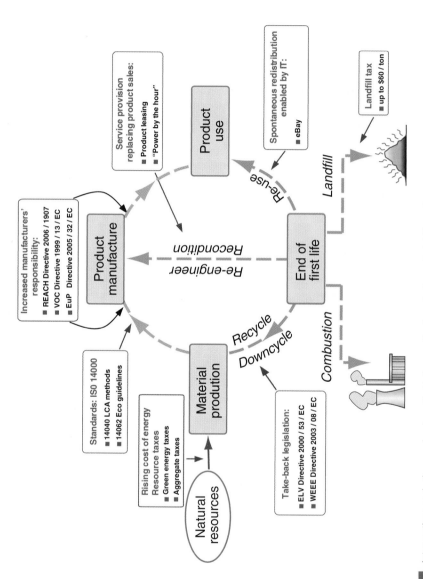

more tools designed to help implement them. The development of tools that integrate with existing product data management (PDM) systems can make compliance semiautomatic, flagging the use of any material that is, in any sense, restricted, and automatically generating the reports that the legislation requires[7].

5.7 Summary and conclusion

Governments intervene when they want to change the way people and organizations behave. Many now accept that the way they behave at present is damaging to the environment in ways that could have catastrophic consequences. Some of this damage is local and can be tackled at a national level by internalizing its cost, making the polluter pay or rewarding those that do not pollute. National and multinational regulations, controls, and directives impose reporting requirements, set tax levels, and establish trading schemes to create incentives for change, with the ultimate aim of making design for the environment a priority.

Some impacts, however, are on a global scale. The externalized costs fall on both the nations that are responsible for the impact and those that are not. Solutions, here, require international agreements. Binding, universal, and enforceable agreements here are out of reach—the diversities of GDP, national ambitions, and political systems are too great. Nonetheless, protocols and statements can be negotiated that many nations feel able to sign. If the IPCC report is to be taken seriously, it is in furthering these international agreements that we must place our hopes for the future.

But the mountain of legislation grows and grows. One only wishes that environmental agencies could aspire to be as concise as Moses.

5.8 Further reading

Ayres, R.U., and L.W. Ayres (Eds.), A handbook of industrial ecology, Edward Elgar, 2002, ISBN 1 84064 506 4. *(Industrial ecology is industrial because it deals with product design, manufacture and use, and ecology because it focuses on the interaction of this process with the environment in which we live. This handbook brings together current thinking on the topic.)*

Brundtland, G.H., Chairman, Our common future, Report of the World Commission on Environment and Development, Oxford University Press, 1987, ISBN 0-19-282080-X. *(Known as the Brundtland Report, it defined the principle of sustainability as "Development that meets the needs of today without compromising the ability of future generations to meet their own needs.")*

[7]See www.emitconsortium.com for an example.

Carson, R., Silent spring, Houghton Mifflin, 1962, republished by Mariner Books, 2002, ISBN 0-618-24906-0. *(Meticulous examination of the consequences of the use of the pesticide DDT and of the impact of technology on the environment.)*

ELV, The Directive EC 2000/53 Directive on End-of-life vehicles (ELV), Journal of the European Communities L269, October 21, 2000, 34–42. *(European Union Directive requiring take-back and recycling of vehicles at end of life.)*

Hardin, G., The tragedy of the commons, Science, 1968, Vol. 162, 1243–1248. *(An elegantly argued exposition of the tendency to exploit a common good such as a shared resource [the atmosphere] or pollution sink [the oceans] until the resource becomes depleted or overpolluted.)*

IPCC, The Intergovernmental Panel on Climate Change, Fourth Assessment Report of the IPCC, UNEP, 2007, www.ipcc.ch. *(The report that establishes beyond any reasonable doubt the correlation between carbon in the atmosphere and climate change.)*

ISO 14040, Environmental management: life-cycle assessment, Principles and framework, 1998.

ISO 14041, Goal and scope definition and inventory analysis, 1998.

ISO 14042, Life-cycle impact assessment, 2000.

ISO 14043, Life-cycle interpretation, International Organization for Standardization, 2000. *(The set of standards defining procedures for life-cycle assessment and its interpretation.)*

Meadows D. H., D. L. Meadows, J. Randers, and W. W. Behrens, The limits to growth, Universe Books, 1972. (The "Club of Rome" report that triggered the first of a sequence of debates in the 20th century on the ultimate limits imposed by resource depletion.)

RoHS, The Directive EC 2002/95/EC on the Restriction of the Use of Certain Hazardous Substances in Electrical and Electronic Equipment, 2002. *(This directive, commonly referred to as the Restriction of Hazardous Substances Directive, or RoHS, was adopted by the European Union in February 2003 and came into force on July 1, 2006.)*

WEEE, The Directive EC 2002/96 on Waste electrical and electronic equipment (WEEE), Journal of the European Communities 37, February 13, 2003, 24–38.

5.9 Exercises

E.5.1. What is a protocol? What do the Montreal Protocol and the Kyoto Protocol commit the signatories to do?

E.5.2. What is meant by *internalized* and *externalized* environmental costs? If an company is required to "internalize its previous externalities," what does it mean?

E.5.3. What is the difference between *command and control* methods and the use of *economic instruments* to protect the environment?

E.5.4. How does emissions trading work?

E.5.5. Carbon trading sounds like the perfect control mechanism to enable emissions reduction. But nothing in this world is perfect. Use the Internet to research the imperfections in the system and report your findings.

E.5.6. What are the merits and difficulties associated with (a) taxation and (b) trading schemes as economic instruments to control pollution?

E.5.7. Your neighbor with a large 4 × 4 boasts that his car, despite its size, is carbon-neutral. What on earth does he mean (or does he think he means)?

E.5.8. In December 2007 Saab posted advertisements urging consumers to "switch to carbon-neutral motoring," claiming that "every Saab is green." In a press release the company said it planned to plant 17 native trees for each car purchased. The company claimed that its purchase of offsets for each car sold made Saab the first car brand to make its entire range carbon-free.

What is misleading about this statement? (The company has since withdrawn it.)

E.5.9. What tools are available to help companies meet the VOC regulations? Carry out a Web search to find out and report your findings.

E.5.10. What tools are available to help companies meet the EuP regulations? Carry out a Web search to find out and report your findings.

Ecodata: values, sources, precision

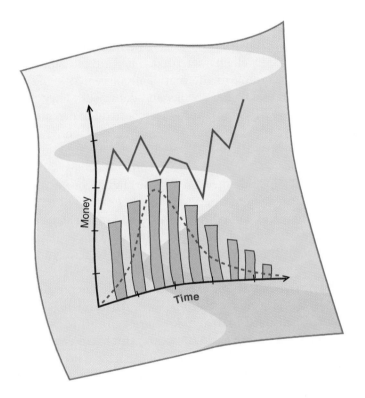

6.1 Introduction and synopsis

Decisions need data. Chapter 12 of this book provides it in the form of a compilation of data sheets for engineering materials. Each has a description and an image and lists values for engineering and ecoproperties. These provide the inputs to the audit and selection methods developed in Chapters 7, 8, and 9. The engineering properties are fully documented in standard texts listed under Further Reading. Those relating to the environment are less familiar, needing further explanation. This chapter describes them and what they mean.

The precision of a great deal of ecodata is low. The values of some are known to within 10%, others with even less certainty. So it is worth first asking the general question: how much precision do you need to deal with a given problem? The answer: just enough to distinguish the viable alternatives. Often that does not require much. As we shall see, precise judgments can be based on imprecise data.

6.2 Data precision: recalibrating expectations

The engineering properties of materials—their mechanical, thermal, and electrical attributes—are well characterized. They are measured with sophisticated equipment according to internationally accepted standards and are reported in widely accessible handbooks and databases. They are not *exact*, but their precision—when it matters—is reported; many are known into three-figure accuracy, some to more. A pedigree like this gives confidence. This is data that can be trusted.

Additional properties are needed to incorporate eco-objectives into the design process. They include measures of the energy committed and carbon released into the atmosphere when a material is extracted or synthesized—its *embodied energy* and *carbon footprint*—and similar data for processing of the material to create a shaped part. There are more properties, introduced in a moment. But before the introductions it helps to know what to expect.

Take embodied energy as an example. It is the energy to produce unit mass (usually, 1 kg) of a material from, well, whatever it is made from. It is a key input to any eco-tool. Unlike the engineering properties, many with a provenance stretching back 200 years, embodied energy is an upstart with a brief and not very creditable history. There are no sophisticated test machines to measure it. International standards, detailed in ISO 14040 and discussed in Chapter 3, lay out procedures, but these are vague and not easily applied. There is no pedigree here; it's a mongrel. So just how far can values for this and other ecoproperties be trusted? An analysis, documented in Section 6.3, suggests a standard deviation of $\pm 10\%$ at best.

Bad news? Not necessarily. It depends on how you plan to use the data. Methods for selecting materials based on environmental criteria must be fit for their purpose. The distinctions they reveal and the decisions drawn from them must be *significant*, meaning that they must stand despite the imprecision of the data on which they are based.

The data sheets of Chapter 12 deal with this issue by listing all properties as ranges: *aluminium: embodied energy 200–240 MJ/kg*, for example. The ranges allow "best-case" and "worst-case" scenarios to be explored. When point (single-valued) data is needed, take the mean of the range.

6.3 The eco-attributes of materials

Table 6.1 lists ecodata for materials in a similar format to the data sheets of Chapter 12. Here we step through the blocks, explaining what the names mean. The data themselves are drawn from many sources. They are listed under Further Reading at the end of this chapter.

Geo-economic data. The first block of data in Table 6.1 contains information about the resource base from which the material is drawn and the rate at which it is being exploited. The *annual world production* is simply the mass of the material extracted annually from ores or feedstock, expressed in terms of the tonnes of metal (or other engineering material) it yields. The *reserve*, as explained in Chapter 2, is the currently reported sizes of the economically–recoverable ores or feedstock from which the material is extracted or created.

Ecoproperties: material production. The *embodied energy H_m* is the energy that must be committed to create 1 kg of usable material—1 kg of steel stock, or of PET pellets, or of cement powder, for example—measured in MJ/kg.

Table 6.1	Eco-attributes of a material
Aluminum alloys	
Geo-economic data for principal component	
Annual world production	33×10^6–34×10^6 tonnes/yr
Reserve	20×10^9–2.2×10^9 tonnes
Ecoproperties: material production	
Embodied energy, primary production	200–240 MJ/kg
CO_2 footprint, primary production	11–13 kg/kg
Water usage	125–375 l/kg
Eco-indicator	740–820 millipoints/kg
Ecoproperties: processing	
Casting energy	2.4–2.9 MJ/kg
Casting CO_2 footprint	0.14–0.17 kg/kg
Deformation processing energy	2.4–2.9 MJ/kg
Deformation processing CO_2 footprint	0.19–0.23 kg/kg
Recycling	
Embodied energy, recycling	18–21 MJ/kg
CO_2 footprint, recycling	1.1–1.2 kg/kg
Recycle fraction in current supply	33–55%

The *CO_2 footprint* is the associated release of CO_2 into the atmosphere, in kg/kg.[1] It is tempting to try to estimate embodied energy via the thermodynamics of the processes involved; extracting aluminum from its oxide, for instance, requires the provision of the free energy of oxidation to liberate it. This much energy must be provided, it is true, but it is only the beginning. The thermodyamic efficiencies of processes are low, seldom reaching 50%. Only part of the output is usable; the scrap fraction ranges from a few percent to more than 10%. The feedstocks used in the extraction or production themselves carry embodied energy. Transport is involved. The production plant itself has to be lit, heated, and serviced. And if it is a dedicated plant, one that is built for the sole purpose of making the material or product, there is an "energy mortgage"—the energy consumed in building the plant in the first place.

Embodied energies are more properly assessed by *input/output analysis*. For a material such as ingot iron, cement powder, or PET granules, the embodied energy/kg is found by monitoring over a fixed period of time the total energy input to the production plant (including that smuggled in, so to speak, as embodied energy of feedstock) and dividing this total by the quantity of usable material shipped out of the plant. Figure 6.1 shows, much simplified, the inputs to a PET production facility: oil derivatives such as naptha and other feedstock, direct power (which, if electric, is generated in part from fossil fuels with a production efficiency of about 35%), and the energy of transporting the materials to the facility.

The material inputs to any processing operation are referred to as *feedstock*. Some are inorganic, some organic. Accounting for inorganic feedstock is straightforward because, during processing, they appear either in the final product or in the waste output. Accounting for organic feedstock is more complex because they can be used either as a material input or as a fuel input. When used as a fuel, the material is burned to provide energy and, once burned, is gone forever. In the case of fossil fuels—oil and gas—their use as fuels represents a depletion of the resource. By contrast, when organics are used as materials (oil as a feedstock for plastics, for example), its energy content is not lost but is rolled up and incorporated into the product. Using it in this way also represents a depletion of the resource, although it does not give rise to any air emissions.

In describing the embodied energy of materials or products it is important to include feedstock energies in the total because they represent a

---

[1] Throughout this book, atmospheric carbon is measured as CO_2 One kg of elemental carbon is equivalent to 3.6 kg of CO_2.

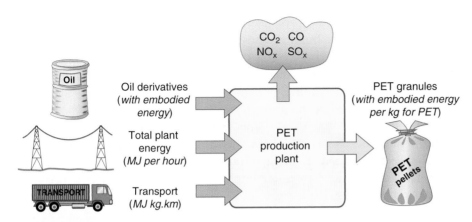

FIGURE 6.1   *The idea of embodied energy. Energy, in various forms, enters or is required by the plant. Its output is a material. The energy per kg of usable material is the embodied energy of the material.*

demand for a resource to support the system of production. The plant of Figure 6.1 has an hourly output of usable PET granules. The embodied energy of the PET, $(H_m)_{PET}$, with usual units of MJ/kg, is then given by

$$(H_m)_{PET} = \frac{\sum Energies\ entering\ plant\ per\ hour}{Mass\ of\ PET\ granules\ produced\ per\ hour}$$

The CO_2 footprint of a material is assessed in a similar way. The carbon emissions consequent on the creation of unit mass of material include those associated with transport, the generation of the electic power used by the plant, and that of feedstocks and hydrocarbon fuels. The CO_2 footprint, with the usual units of kg of CO_2/kg, is then the sum of all the contributions per unit mass of usable material exiting the plant.

Plants grow by absorbing CO_2 from the atmosphere and H_2O from the Earth to build hydrocarbons, notably cellulose and lignin, the building blocks of wood. Wood thus sequesters carbon. This has led to the statement that the carbon footprint of wood is negative; wood absorbs carbon rather than releasing it.

Many ecostatements need close scrutiny and this is one. Coal is a hydrocarbon, derived, like wood, from plant life. The carbon in coal was once in the atmosphere; the formation of coal sequestered it. But we do not credit coal with a negative carbon footprint because, when we use it, we do not replace it; the carbon it contains is returned to the atmosphere. A credit is only real if the resource is replaced.

Some plant life, useful in engineering, can be grown "sustainably." Hemp, used in rope and fabrics, building construction, and, increasingly, to reinforce polymer composites, is one. Hemp can be grown without fertilizers, and it grows fast; it can be grown as fast as it is used. The carbon it sequesters is retained in the fabric, building, or composite made from it, so its use generally removes carbon from the atmosphere, returning it when it is burned or decomposes at end of life. A carbon-neutral cycle is possible here.

So the question: is wood like coal or is it like hemp? At an international level, the world's forests are being cut down for construction, fuel, pulping, and mere clearing at a far greater rate than they are replanted. Trees take 80 years to grow to maturity; planting trees to offset the carbon footprint of driving a car (average life: 14 years) does not add up. Until stocks are replaced as fast as they are used, wood has to be viewed as a resource that is more like coal than hemp, with a corresponding positive carbon footprint. The data sheets for wood and wood-based materials in Chapter 12 take this approach.

Wood-based products—plywood, chipboard, fiberboard and the like—have the merit that they use more of the trunk than solid wood beams or panelling do, and they can be made from lower-grade timber. But all involve a significant component of a polymer adhesive, driving the energy content and carbon footprint yet higher.

Data precision. Figure 6.2 plots some of the reported values of embodied energy for aluminum over time, starting with the earliest measurements, around 1970. The mean value of the data plotted here is 204 MJ/kg. But the standard deviation is 58 MJ/kg, and that is 25% of the mean. A closer examination of the data sources shows some to be more rigorous than others, so a more selective plot shifts the mean to 220 MJ/kg and reduces the standard deviation to, perhaps, 10%. Hammond and Jones (2006) plot embodied energies for a number of materials in this way. They all have the same feature: a standard deviation of between 10% and 50% of the mean. It is worth remembering that engineers don't design on a *mean* property value but on an "allowable" that is somewhere between 1 and 6 standard deviations on the safe side of the mean. So, the bottom line—when dealing with embodied energies such as the 220 MJ/kg quoted previously, read them thus: the first figure can be trusted; the second is debatable; the third is *meaningless*.

Water usage. The data sheets list the approximate quantity of commercial water, in liters, used in the production of 1 kg of material.

Eco-indicators. As explained in Chapter 3, attempts have been made to roll up the energy, water, and gaseous, liquid, and solid emissions associated

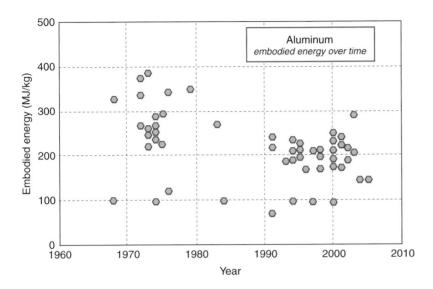

FIGURE 6.2 *Data for the embodied energy of aluminum. The mean is 204 MJ/kg, with a standard deviation of 58 MJ/kg. Using the best-characterized data only gives a mean of 220 MJ/kg with a standard deviation of 20 MJ/kg.*

with material production into a single number, or *eco-indicator*. The records in Chapter 12 include the value when it is available.

Ecoproperties: processing energy and CO_2 footprint. Product manufacture requires that materials be shaped, joined, and finished. The processing energy H_p associated with a material is the energy, in MJ/kg, used to do this. Thus polymers, typically, are molded or extruded; metals are cast, forged, or machined; ceramics are shaped by powder methods, composites by molding or lay-up methods (see Table 6.2). A characteristic energy per kg is associated with each of these. The energy consumed by a casting furnace or an injection-molding machine can be directly measured, but the production plant as a whole uses more than this through the provision of transport, heating, lighting, management, and maintenance.

A more realistic measure of processing energy is the total energy entering the plant (but excluding that embodied in the materials that are being shaped since that is attributed to the material production phase) divided by the weight of usable shaped parts that it delivers, giving a value in MJ/kg. Continuing with the PET example, the granules now become the input (after transportation) to a facility for blow-molding PET bottles for water, as shown in Figure 6.3. As before, energy and feedstock are consumed, emissions and waste are excreted, and the products—PET blow-molded bottles—emerge.

Table 6.2	Shaping methods

Shaping of metals

Casting
Forging, rolling (deformation processing)
Metal powder processing
Vapor-phase methods

Shaping of polymers

Polymer molding (thermo-forming, injection molding, etc.)
Polymer extrusion
Polymer machining

Shaping of ceramics

Ceramic powder forming

Shaping of glasses

Glass molding

Shaping of composites

Simple composite molding
Advanced composite molding

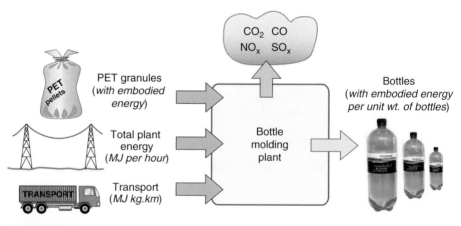

FIGURE 6.3 *Manufacture: here, the blow-molding of PET bottles.*

The output of the analysis is the energy-committed per unit weight of bottles produced. The CO_2 footprint is evaluated in a similar way.

The availability and precision of data for processing are particularly poor due to differences in processing equipment and manufacturing practice. Reports of the total annual energy consumed by the metal-casting and

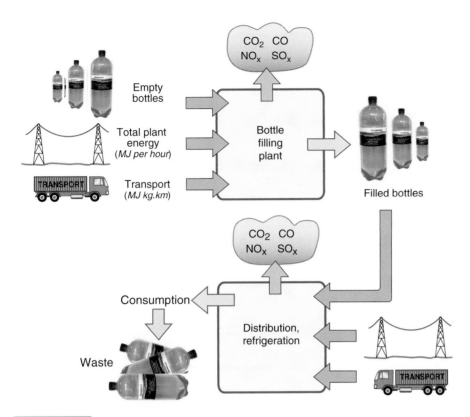

FIGURE 6.4 *The use phase of PET water bottles: filling, distribution, and refrigeration. Energy is consumed in transport and refrigeration.*

polymer-molding industries in the United States are available; these also list the annual tonnage of usable castings and moldings they produce. Information of this sort allows estimates of processing energies for a number of primary shaping processes. It is these that appear in the records of Chapter 12.

Processing energies are generally smaller than the energy embodied in the material or that consumed during its use. This being so, the lack of precision here is less critical than if the reverse were true.

Recycling and end-of-life. The product is now transported to the point of sale, passes to the consumer, is used, and, ultimately, reaches the end of its useful life (see Figure 6.4).

Underlying data for transport and use are explored in Section 6.4. Here we jump to the last block of attributes listed in Table 6.1: those relating to *recycling*. The existence of embodied energy has another consequence: that the energy to recycle a material is sometimes much less than that required

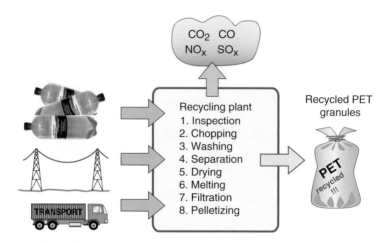

FIGURE 6.5 *Recycling. Many steps are involved, all of which consume some energy, but the embodied energy of the material is conserved.*

Table 6.3	Precious metals (catalysts, electrodes, contacts)	
Metal	**Embodied energy (MJ/kg)**	**Carbon footprint (kg/kg)**
Silver	1.7×10^3–2.0×10^3	100–115
Palladium	34×10^3–49×10^3	2140–2360
Gold	68×10^3–72×10^3	4100–4500
Platinum	114×10^3–120×10^3	8740–8660

for its first production because the embodied energy is retained (see Figure 6.5). The recycling energy, listed in the data sheets, is, despite its approximate nature, a useful indicator of the viability of recycling. Typical values lie in the range of 10–100 MJ/kg.

Special cases: precious metals and electronics. Precious metals are used in small quantities, but they carry a heavy burden of embodied energy, carbon, and, of course, cost. But they have unique properties: exceptional electrical conductivity, exceptional resistance to corrosion, and exceptional properties as electrodes, sensors, and catalysts. A rule of convenience in energy accounting, used later, is to ignore (or approximate) the contribution of parts that contribute less than 5% of the weight of the product, but we can't do that with precious metals; even a small quantity adds significantly to the energy and carbon totals. Table 6.3 lists the data needed to include them.

Table 6.4	Approximate embodied energies and CO_2 footprints for electronic components	
Component	**Embodied energy (MJ)**	**Carbon footprint (kg)**
Small (handheld) electronic devices, depending on scale *(per kg)*	2000–4000	200–400
Displays, depending on size *(per m²)*	2950–3750	295–375
Assembling of printed wiring boards *(per kg)*	120–140	12–14
Batteries (Ni-Cd rechargeable) *(per kg)*	180–220	18–22

The same is true of electronic components: integrated circuits, surface-mount devices, displays, batteries, and the like. They are now an integral part of almost all appliances. They don't weigh much, but they are energy-intensive to make. Table 6.4 lists sufficient data to include them at an approximate level.

6.4 Energy and CO_2 footprints of energy, transport, and use

Energy is used to make materials and to shape, join, and finish them to make products. Energy is used to transport the products from where they are made to where they are used, and the products themselves use energy during their lifetimes; some use a great deal. This energy is provided predominantly by fossil fuels and by electric power, much if it generated from fossil fuels. These sources differ in their energy intensity and carbon release.

Energy intensities. Energy intensities of fossil fuels and their carbon footprints are listed in Table 6.5. Reading across, there is the fuel type, the oil equivalent (OE, the kg of crude oil with the same energy content as 1 kg of the fuel), the energy content per unit volume and per unit weight, and the CO_2 release per unit volume, weight, and MJ. The units (kg, KJ, etc.) used here are those standard to the SI system. Conversion factors to other systems (lbs, Btu, etc.) can be found at the end of this book.

The oil equivalence of electric power. Electricity is the most convenient form of energy. Today most electricity is still generated by burning fossil fuels, but the pressure on these fuels and the problems caused by the emissions they release are urging governments to switch to nuclear and renewable sources, and most of these generate electric power. The *energy mix* in a country's electricity supply is the proportional contribution of each source

Table 6.5	The energy intensity of fossil fuels and their carbon footprints					
Fuel	kg OE*	MJ/liter	MJ/kg	CO_2, kg/liter	CO_2, kg/MJ	CO_2, kg/kg
Coal, lignite	0.45	—	18–22	—	0.080	1.6
Coal, anthracite	0.72	—	30–34	—	0.088	2.9
Crude oil	1.0	38	44	3.1	0.070	3.0
Diesel	1.0	38	44	3.1	0.071	3.2
Gasoline	1.05	35	45	2.9	0.065	2.89
Kerosene	1.0	35	43.8	3.0	0.068	3.0
Ethanol	0.71	23	31	2.8	0.083	2.6
Liquid natural gas	1.2	25	55	3.03	0.055	3.03

*Kilograms oil equivalent (the kg of oil with the same energy content)

Table 6.6	Electricity generation, energy mix, MJ oil per kW.hr, and CO_2 per kW.hr					
Country	Fossil fuel (%)	Nuclear (%)	Renewables (%)	Efficiency[a] (%)	MJ OE[b] per kW.hr[d]	CO_2[c], kg per kW.hr[d]
Australia	92	0	8	33	10.0	0.71
China	83	2	15	32	9.3	0.66
France	10	78	12	40	0.9	0.06
India	81	2.5	16.5	27	10.8	0.77
Japan	61	27	12	41	5.4	0.38
Norway	1	0	99	—	0	0
United Kingdom	75	19	6	40	6.6	0.47
United States	71	19	10	36	7.1	0.54

[a] Conversion efficiency of fossil fuel to electricity.

[b] MJ of fossil fuel (oil equivalent) used in energy mix per kW.hr of delivered electricity from all sources.

[c] CO_2 release per kW.hr of delivered electricity from all sources.

[d] 1 kW.hr is 3.6 MJ.

to the total. The first four columns of Table 6.6 give examples of this mix for countries that span the extremes. Australia relies on fossil fuels for almost all its electricity; in France it is predominantly nuclear, and in Norway it is almost wholly hydroelectric. The United States and Japan are the world's two largest economies, but the two most populous and fastest growing are China and India.

The relevant numbers from an environmental point of view are those in the last three columns: the efficiency of electricity generation from fossil fuels and the oil equivalence and CO_2 release per unit of delivered electrical power. These figures differ greatly from country to country, mainly because of the differing energy mixes, to a lesser extent because of the differing efficiencies. We'll use these numbers in later chapters. To keep things simple we will use values for a "typical" developed country with an energy mix of 75% fossil fuel and a conversion efficiency of 38%, giving an oil equivalence of 7 MJ (or 0.16 kg) and a carbon footprint of 0.5 kg CO_2 per kW.hr.

Transport. Manufacturing is now globalized. Products are made where it is cheapest to do so and then transported, frequently over large distances, to the point of sale. Transport is an energy-conversion process: primary energy (oil, gas, coal) is converted into mechanical power and thus motion, sometimes with an intermediate conversion to electrical power. As in any energy conversion process, there are losses. We express the energy of transport as an energy per tonne.km, carrying with it an associated CO_2 footprint per tonne.km.

Table 6.7 lists data for these quantities, drawn from reports of the U.S. Department of Transport, the U.K. Network Rail, and Transport Watch U.K. Transport energy and carbon footprint are calculated by multiplying the weight of the product by the distance traveled and the fuel-vehicle coefficients listed in the table.

Use energy. Many products consume energy, or energy is consumed on their behalf, during the use phase of life. As we shall see in Chapter 7, this use-phase energy is often larger than that of any other. Most of it derives from fossil fuels, and some is consumed in that state, as primary fossil fuel. Much is first converted to other forms of energy before it is used, of which the most obvious is electricity.

When fossil fuels are used directly (as in the use of gasoline to power cars), the primary energy and CO_2 can be read directly from Table 6.5. When instead it is used as electricity, relevant fossil fuel energy and CO_2 depend on the energy mix and generation efficiency, and these differ from country to country. It is then necessary to convert the electrical energy, usually given in kW.hr, to MJ and CO_2 oil equivalent by multiplying by the conversion factors like those of Table 6.6.

Table 6.7	The energy and CO_2 costs of transport	
Fuel and vehicle type	**Energy (MJ/tonne.km)**	**Carbon emission (kg CO_2/tonne.km)**
Diesel—ocean shipping	0.16	0.015
Diesel—coastal shipping	0.27	0.019
Diesel—rail	0.31	0.022
Diesel—32 tonne truck	0.46	0.033
Diesel—14 tonne truck	0.90	0.064
Diesel—light goods vehicle	1.36	0.097
Gasoline—family car	2.06	0.14
Diesel—family car	1.60	0.11
LPG—family car	3.87	0.18
Gasoline—hybrid family car	1.55	0.10
Gasoline—super-sports and SUV	4.76	0.31
Kerosene—long-haul aircraft	8.30	0.55
Kerosene—short-haul aircraft	15.0	1.00
Kerosene—helicopter (Eurocopter AS 350)	55.0	3.30

6.5 Exploring the data: property charts

Data sheets like those of Chapter 12 list material properties, but they present no comparisons and they give no perspective. The way to achieve these is to plot *material property charts*.

Material property charts. Property charts are of two types: *bar charts* and *bubble charts*. A bar chart is simply a plot of the value ranges of one property. Figure 6.6 shows an example: it is a bar chart for Young's modulus, *E*, the mechanical property that measures stiffness. The largest value is more than 10 million times greater than the smallest—many other properties have similarly large ranges—so it makes sense to plot them on logarithmic[2] scales (as here), not linear ones. The length of each bar shows the range of the property for each of the materials, here segregated by family. The differences between the families now become apparent. Metals and ceramics

---

[2] *Logarithmic* means that the scale goes up in constant multiples, usually of 10. We live in a logarithmic world; our senses, for instance, all respond in that way.

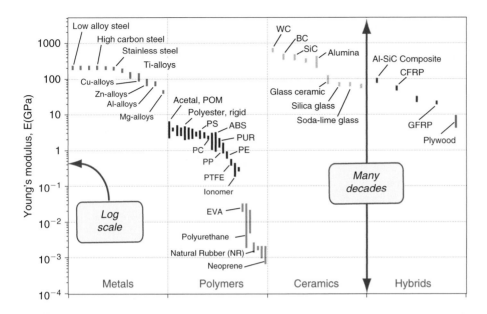

FIGURE 6.6 | *A bar chart of modulus. It reveals the difference in stiffness between the families.*

have high moduli. Those of polymers are smaller by a factor of about 50 than those of metals. Those of elastomers are some 500 times smaller still.

More information is packed into the picture if two properties are plotted to give a *bubble chart*, as in Figure 6.7, here showing modulus E and density ρ. As before, the scales are logarithmic. Now families are more distinctly separated. Ceramics lie in the yellow envelope at the very top; they have moduli as high as 1000 GPa. Metals lie in the reddish zone near the top right; they, too, have high moduli, but they are heavy. Polymers lie in the dark blue envelope in the center, elastomers in the lighter blue envelope below, with moduli as low as 0.0001 GPa. Materials with a lower density than polymers are porous—manmade foams and natural cellular structures like wood and cork. Each family occupies a distinct, characteristic field. Yet more information can be displayed by using functions of properties—groupings such as E/ρ or $H_m\rho$ for the axes of the charts; examples appear later.

Material property charts are a core tool.[3]

- They give an overview of the physical, mechanical, and functional properties of materials, presenting the information about them in an compact way.

---

[3] Further descriptions and a wide range of charts can be found in Ashby (2005) and Ashby et al. (2006).

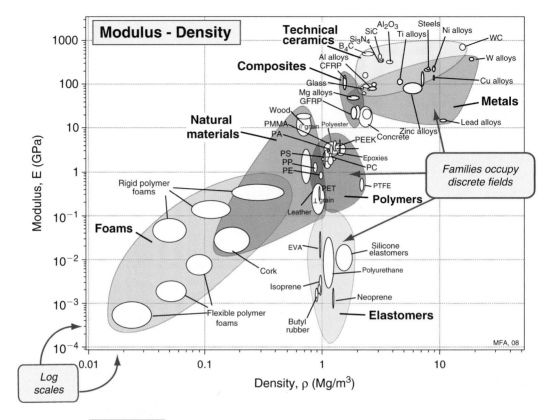

FIGURE 6.7 *A bubble chart of modulus and density. Families occupy discrete areas of the chart.*

- They reveal aspects of the physical origins of properties, helpful in understanding the underlying science.

- They become a tool for optimized selection of materials to meet given design requirements, and they help us understand the use of materials in existing products.

- They allow the properties of new materials, such as those with nano or amorphous structures, to be displayed and compared with those of conventional materials, bringing out their novel characteristics and suggesting possible applications.

Property charts appear in the chapters that follow. Right now we use them to explore the ecodata.

Embodied energies of materials. Embodied energies of materials are compared in the bar charts of Figures 6.8 and 6.9. The first plots the energy

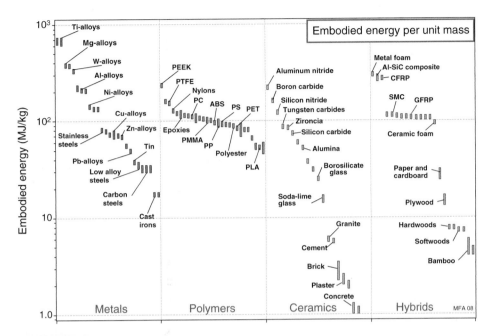

FIGURE 6.8 *A bar chart of the embodied energies of materials per unit mass.*

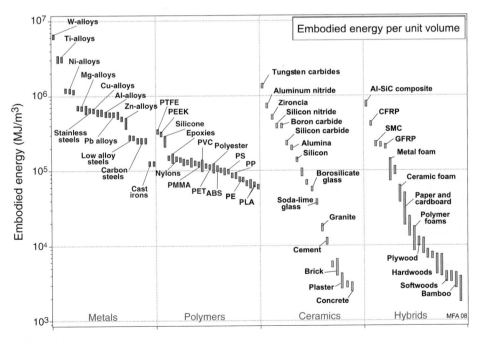

FIGURE 6.9 *A bar chart of the embodied energies of materials per unit volume.*

per unit mass (units: mJ/kg). Among metals, the light alloys based on aluminum, magnesium, and titanium have the highest values, approaching 1000 MJ/kg for titanium on this chart, but precious metals lie much higher still (Table 6.3). Polymers all cluster around 100 MJ/kg, less than the light alloys but considerably more than steels and cast irons, with energies between 20 MJ/kg and 40 MJ/kg. Technical ceramics such as aluminum nitride have high energies; those for glass, cement, brick, and concrete are much lower. Composites, too, have a wide spread. High-performance composites—here we think of carbon-fiber reinforced polymers (CFRPs) —lie at the top, well above most metals. At the other extreme, paper, plywood, and timber are comparable with the other materials of the construction industry.

But is embodied energy *per unit mass* the proper basis of comparison? Suppose, instead, the comparison is made *per unit volume* (see Figure 6.9). The picture changes. Now metals as a family lie above the others. Polymers cluster around a value that is lower than most metals; by this measure they are not the energy-hungry materials they are sometimes made out to be. The nonmetallic materials of construction—concrete, brick, wood—lie far below all of them. CFRP is now comparable with aluminum.

This raises an obvious question: if we are to choose materials with minimum embodied energy as an objective, what basis of comparison should we use? A mistaken choice invalidates the comparison, as we have just seen. The right answer is to compare energy *per unit of function*. We return to this topic, in depth, in later chapters.

Carbon footprint. Material production pumps enormous quantities of CO_2 into the atmosphere; some 20% of the global total arises in this way. So it is interesting to ask: which materials contribute the most? That depends on the carbon footprint per kg and on the number of kgs per year that are produced. The data sheets have the information to explore this question. Figure 6.10 answers the question and illustrates how the data can be used. It was made by multiplying the annual world production by the carbon footprint for material production to give the tonnage of CO_2 per material per year. The big four are iron and steel, aluminium, concrete (cement), and paper and cardboard. They account for much more than all the rest put together.

Water usage. Water usage is compared in Figure 6.11. Materials with high embodied energy tend to have high water usage—not surprising, given the water demands of energy listed in Table 2.2. There is not much else to be said except that the water consumptions plotted here are small compared with those required, per kg, for water-intensive agricultural crops such as rice and cotton or for materials derived from animal husbandry, such as wool.

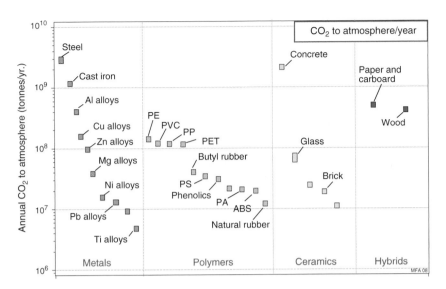

FIGURE 6.10   *Annual carbon dioxide to atmosphere from material production.*

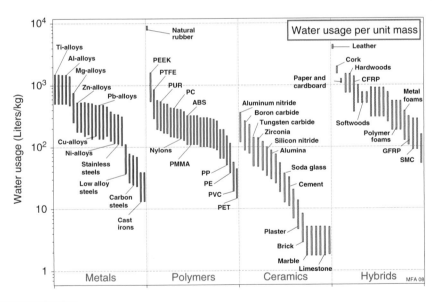

FIGURE 6.11   *Water usage bar chart.*

Process energies. Process energies are plotted in Figure 6.12. Here it is necessary to remember the very limited and imprecise nature of the data. Vapor processing and powder methods stand out as energy intensive.

Recycling. Figure 6.13 presents the data for recycle fraction in current supply. As discussed in Chapter 4, the recycling of metals is highly developed

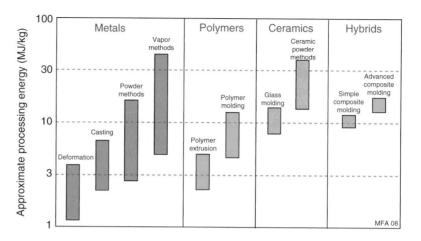

FIGURE 6.12   *Approximate processing energies for materials.*

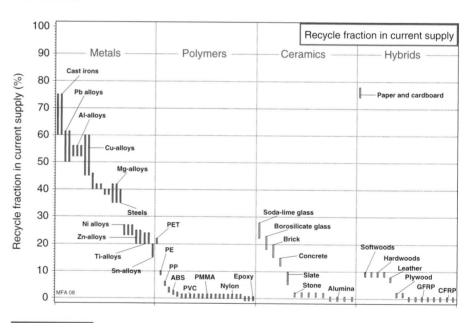

FIGURE 6.13   *Recycle fraction bar chart.*

and its contribution to current supply is large. The same cannot be said of polymers. The commodity polymers are used in large quantities, many in products with short life, and they present major problems in waste management, all of which, you would think, would encourage effective recycling. But the economics of polymer recycling are unattractive, with the result that their contribution to current supply is small.

6.6 Summary and conclusion

You can't answer technical questions without numbers. "Choice X is worse than Choice Y ..." is a statement that is on solid ground only if you have data to demonstrate that it is indeed so. Concern for the environment today sometimes leads to statements based more on emotion than reason, clouding issues and breeding deception. So, boring though numbers can be, they are essential for analysis based on fact, not on speculation.

To use numbers, you have to know what they mean and how accurate (or inaccurate) they are. This chapter introduced the ones used later in this book, presenting them as bar charts that display relationships and correlations. The numbers describe eco-attributes of materials, here principally those relating to energy and carbon footprint. The thing to remember about them is that their precision is low. If you are going to base decisions on their values, make sure that the decision still stands if the numbers are wrong by ±10%, or, better, ±20%. Such uncertainty does not prevent decision making, provided that its presence is recognized and allowed for.

6.7 Further reading

The references are segregated by data type following the section headings in this chapter.

General engineering properties of materials

Ashby, M.F., Shercliff, H.R. and Cebon, D. (2007), "Materials: engineering, science, processing and design", Butterworth Heinemann. ISBN-13: 978-0-7506-8391-3. (*An elementary text introducing materials through material property charts.*)

Callister, W.D. (2003), "Materials science and engineering, an introduction", 6th ed., John Wiley. ISBN 0-471-13576-3. (*A well-respected materials text, now in its 6th edition, widely used for materials teaching in North America.*)

Charles, J.A., Crane, F.A.A. and Furness, J.A.G. (1997), "Selection and use of engineering materials", 3rd ed., Butterworth Heinemann. ISBN 0-7506-3277-1. (*A materials science approach to the selection of materials.*)

Dieter, G.E. (1991), "Engineering design, a materials and processing approach", 2nd ed., McGraw-Hill. ISBN 0-07-100829-2. (*A well-balanced and respected text focusing on the place of materials and processing in technical design.*)

Farag, M.M. (1989), "Selection of materials and manufacturing processes for engineering design", Prentice-Hall. ISBN 0-13-575192-6. (*Like Charles, Crane, and Furness, this text presents a materials science approach to the selection of materials.*)

Kalpakjian, S. and Schmid, S.R. (2003), "Manufacturing processes for engineering materials", 4th ed., Prentice Hall, Pearson Education. ISBN 0-13-040871-9. (*A comprehensive and widely used text on material processing.*)

Geo-economic data

Chemsystems, 2006, www.chemsystems.com. *(Data for polymers.)*

Cheresources, 2007, www.cheresources.com/polystyz5.shtml. *(Data for polymers.)*

Geokem, 2007, www.geokem.com/global-element-dist1.html. *(Data for metals and minerals.)*

International Rubber Study Group (IRSG), Vol. 61, Nos. 4 and 5, January/February 2007. *(Data for rubber.)*

Pilkington Group Ltd., Merseyside, UK, 2007. *(Data for glass.)*

Pulp and Paper International, www.jpa.gr.jp/en/about/abo/d.html. *(Data for paper.)*

U.S. Geological Survey, 2007, http://minerals.usgs.gov/minerals/pubs/commodity/. *(Data for metals and minerals.)*

Material production: embodied energy and CO_2, engineering materials

Aggregain (2007), "The Waste and Resources Action Program (WRAP)", www.wrap.org.UK, ISBN 1-84405-268-0. *(Data and an Excel-based tool to calculate energy and carbon footprint of recycled road-bed materials.)*

AMC (2006), "Australian Magnesium Corporation", www.aph.gov.au/house/committee/environ/greenhse/gasrpt/Sub65-dk.pdf.

APME, Association of Plastics Manufacturers in Europe, Eco profiles of the European plastics industry, Brussels, Belgium, 1997,1998, 1999, 2000, www.lca.apme.org.

BCA (2007), "A carbon strategy for the cement industry", British Cement Association, 2007, www.cementindustry.co.uk.

Boustead Model 5, Boustead Consulting, West Sussex, UK, 2007, www.boustead-consulting.co.uk. *(An established life-cycle assessment tool.)*

Boustead, I., APME Association of Plastics Manufacturers in Europe, Report series, Building Research Establishment (2006) BRE Environmental Profiles database, BRE Environment Division, BREEM Center, UK, 1999–2006.

BUWAL, Bundesamt für Umwelt, Wald und Landwirtschaft Environmental Series No 250, Life Cycle Inventories for Packaging. Volume I and II, 1996.

Chapman, P.F. and Roberts, F. (1993), "Metals resources and energy", Butterworths.

Chemlink Australasia, 1997, www.chemlink.com.au/mag&oxide.htm.

Energy Information Association (2008), www.eia.doe.gov. *(Official energy statistics from the U.S. Government.)*

ELCD (2008), http://lca.jrc.ec.europa.eu. *(A high-quality life-cycle inventory, or LCI, core data set of this first version of the Commission's European Reference Life Cycle Data System, or ELCD.)*

European Aluminum Association, 2000, www.eaa.net.

GREET, Argonne National Laboratory and the US Department of Transport, 2007, www.transportation.anl.gov/. *(Software for analyzing vehicle energy use and emissions.)*

Hammond, G. and Jones, C. (2006), "Inventory of Carbon and Energy (ICE), Department of Mechanical Engineering", University of Bath.

Hill, T. (2007), "Civil Engineering Division", Cardiff University School of Engineering. http://carlos.engi.cf.ac.uk/

International Aluminum Institute, Life-cycle inventory of the worldwide aluminum industry, Part 1: automotive, 2000, www.world-aluminum.org.

Kemna, R., van Elburg, M., Li, W. and van Holsteijn, R. (2005), "Methodology study eco-design of energy-using products", Van Holsteijn en Kemna BV (VHK).

Kennedy, J. (1997), "Energy minimization in road construction and maintenance", a Best Practice Report for the Department of the Environment, UK.

Lafarge Cement, Lafarge Cement UK.

Lawson, B. (1996), "Building materials, energy and the environment: towards ecologically sustainable development", RAIA, Canberra. www.greenhouse.gov.au/yourhome/technical/fs31.htm.

Lime Technology (2007), www.limetechnology.co.uk/whylime.

MEEUP Methodology Report, final, VHK, 2005, www.pre.nl/EUP/. (*A report by the Dutch consultancy VHK commissioned by the European Union, detailing their implementation of an LCA tool designed to meet the EU Energy-using Products directive.*)

Ohio Department of Natural Resources, Division of Recycling (2005), www.dnr.state.oh.us/recycling/awareness/facts/tires/rubberrecycling.htm.

Pilz, H., Schweighofer, J. and Kletzer, E. (2005), "The contribution of plastic products to resource efficiency, Gesellschaft fur umfassende analysen (GUA)", Vienna.

Schlesinger, M.E. (2007), "Aluminum recycling", CRC Press. ISBN 0-8493-9662-X.

Stiller, H. (1999), "Material intensity of advanced composite materials", Wuppertal Institut fur Klima, Umvelt, Energie. ISSN 0949-5266.

Sustainable concrete (2008), www.sustainableconcrete.org.uk/. (*A Website representing U.K. concrete producers, carrying useful information about carbon footprints.*)

Szargut, J., Morris, D.R. and Steward, F.R. (1988), "Energy analysis of thermal, chemical and metallurgical processes", Hemisphere.

Szokolay, S.V. (1980), "Environmental science handbook: for architects and builders", Construction Press.

The Nickel Institute North America, Nickel Institute, Toronto, 2007, www.nickelinstitute.org.

Waste Online (2008), www.wasteonline.org.uk/resources/InformationSheets/Plastics.htm.

Material production: embodied energy and CO_2, precious metals

GREET, Argonne National Laboratory and the US Department of Transport (2007), www.transportation.anl.gov/. (*Software for analyzing vehicle energy use and emissions.*)

London Platinum and Palladium Market (2006), www.lppm.org.uk.

Lonmin Plc., 2005 Corporate Accountability Report, www.lonmin.com.

TEAM, Tool for Environmental Analysis and Management (2008), www.ecobalance.com. (*TEAM is Ecobilan's Life Cycle Assessment software. It allows the user to build and use a large database and to model systems associated with products and processes following the ISO 14040 series of standards.)*

Embodied processes and energies, water requirements for electronic components

Kuehr, R. and Williams, E. (Eds.) (2003), "Computers and the environment: understanding and managing their impacts", Kluwer Academic Publishers and United Nations University. ISBN 1-4020-1679-4. *(A multiauthor monograph on the environmental aspects of electronic devices, with emphasis on the WEEE regulations relating to end of life. Chapter 3 deals with the environmental impacts of the production of personal computers.)*

MEEUP, Methodology study eco-design of energy-using products, Final report, VHK, Delft, Netherlands, and the European Commission, Brussels, 2006. *(A study commissioned by the European Union into the development of software to meet the Energy-using Product Directive.)*

Water

AZoM, A to Z of Materials Journal of Materials Online, 2008, www.azom.com.

Chiang, S.H. and Moeslein, D. (1978–1980), "Analysis of water use in nine industries, Parts 1–9, Department of Chemical and Petroleum Engineering", University of Pittsburgh.

Davis, J. R., in ASM Specialty Handbook, ASM International, 1995.

Hill, Water use in industries of the future, Chapter 2, U.S. Department of Energy, July 2003.

Implicit price deflators in national currency and U.S. dollars, United Nations Statistics Division, 2006, http://unstats.un.org/unsd/snaama/dnllist.asp.

Lenzena, M. (2001), "An input-output analysis of Australian water usage", Water Policy, Vol. 3, pp. 321–340.

Leontief, W. (1970), "Environmental repercussions and the economical structure: an input-output approach", The review of Economics and Statistics, Vol. 52, No. 3, pp. 262–271.

Pearce, F., Earth: the parched planet, New Scientist, February 2006, Vol. 2540

Proops, J.L.R. Input-output analysis and energy intensities: a comparison of some methodologies, Appl. Math. Modelling, March 1997, Vol. 1

UNESCO, World water assessment programme, United Nations Educational, Scientific, and Cultural Organization, 2006, www.unesco.org/water.

Vela'zquez, T.E. (2006), "An input-output model of water consumption: analyzing intersectoral water relationships in Andalusia", Ecological Economics, Vol. 56, pp. 226–240.

Aggregated measures: eco-indicators

EPS, Life cycle analysis in product engineering, Environmental Report 49, Volvo Car Corp., 1993.

Goedkoop, M., S. Effting, and M. Collignon, The Eco-indicator 99: a damage-oriented method for life cycle impact assessment, Manual for Designers, April 14, 2000, www.pre.nl.

Idemat Software version 1.0.1, Faculty of Industrial Design Engineering, Delft University of Technology, 1998.

Material processing: energy and CO_2

Allen, D.K. and Alting, L. (1986) "Manufacturing processes", Brigham Young University.

Boustead Model 4, Boustead Consulting, 1999, www.boustead-consulting.co.uk.

Lawrence Berkeley National Laboratory report, Energy use and carbon dioxide emissions from steel production in China, LBNL-47205, April 2005.

Reduced Energy Consumption in Plastics Engineering, from the 2005 European Benchmarking Survey of Energy Consumption, www.eurecipe.com (Eurecipe 05).

The Cast Metal Coalition, 2005, http://cmc.aticorp.org/. *(Energy data based on U.S. national figures.)*

Thiriez, A., An environmental analysis of injection molding, Masters thesis, Massachusetts Institute of Technology, 2006.

U.S. Department of Energy, 1997, www.doe.gov.

U.S. Environmental Protection Agency, 1995, www.epa.gov.

U.S. Department of Energy (2008), "Supporting industries energy and environmental profile" (www1.eere.energy.gov/industry/energy_systems/pdfs/si_profile.pdf).

Recycling and end of life

Aggregain, The Waste and Resources Action Program (WRAP) (2007), ISBN 1-84405-268-0, www.wrap.org.UK.

AMC, Australian Magnesium Corporation (2006), www.aph.gov.au.

Chemlink Australasia (1997), www.chemlink.com.au.

Geokem (2007), www.geokem.com/global-element-dist1.html.

Hammond, G. and Jones, C. (2006), "Inventory of carbon and energy (ICE), Dept. of Mechanical Engineering", University of Bath.

Hill, T. (2007), "Civil Engineering Division", Cardiff University School of Engineering. http://carlos.engi.cf.ac.uk.

International Aluminum Institute (2000), Life cycle inventory of the worldwide aluminum industry: part 1, automotive. www.world-aluminum.org.

Kemna, R., van Elburg, M., Li, W. and van Holsteijn, R. (2005), "Methodology study eco-design for energy-using products", Van Holsteijn en Kemna BV (VHK).

Lafarge Cement UK, 2007.

Lawson, B. (1996), "Building materials, energy and the environment", RAIA, Canberra. www.greenhouse.gov.au.

Ohio Department of Natural Resources, Division of Recycling (2005), www.dnr.state.oh.us.

Pilz, H., Schweighofer, J. and Kletzer, E. (2005), "The contribution of plastic products to resource efficiency, Gesellschaft für umfassende analysen (GUA)", Vienna.

Schlesinger, M.E. (2007), "Aluminum recycling", CRC Press. ISBN 0-8493-9662-X.

Sustainable Concrete, 2008, www.sustainableconcrete.org.uk.

The Nickel Institute North America, 2007, www.nickelinstitute.org.

U.S. Environmental Agency, 2007, www.eia.doe.gov.

U.S. Geological Survey, 2007, http://minerals.usgs.gov.

Waste Online, 2007, www.wasteonline.org.uk.

Transport and use energies

AggRegain, CO_2 emissions estimator tool, published by WRAP (Waste & Resources Action Programme), U.K., 2006, www.aggregain.org.uk.

Battery-powered cars: see http://en.wikipedia.org/wiki/Battery_electric_vehicle.

Carbon Trust, Carbon footprint in the supply chain, 2007, www.carbontrust.co.uk.

Congressional Budget Office, U.S. Congress Energy use in freight transportation, Washington, D.C., 1982.

International Chamber of Shipping, International Shipping Federation, Annual review 2005, www.marisec.org.

Manicore, 2008, www.manicore.com. (*Useful discussion definition of oil equivalence of energy sources.*)

Network Rail (2007) www.networkrail.co.uk/ (*The web site of the UK rail track provider.*)

Shell Petroleum (2007) "How the energy industry works", Silverstone Communications Ltd., ISBN978-0-9555409-0-5.

Transport Watch UK, (2007) www.Transwatch.co.uk.

TRL UPR, Energy consumption in road construction and use, Transport Research Laboratory, UK, 1995.

U.S. Department of Transportation Maritime Administration, Environmental advantages of inland barge transportation, 1994.

Fuel mix in electrical energy

Boustead Model 5, Boustead Consulting, 2007, www.boustead-consulting.co.uk. (*An established life-cycle assessment tool.*)

ELCD, (2008) http://lca.jrc.ec.europa.eu. (*A high-quality life-cycle inventory, or LCI, core data set of this first version of the Commission's European Reference Life Cycle Data System, or ELCD.*)

Electricity Information, IEA Publications, 2008, ISBN 978-9264-04252-0. (*One of a series of IEA statistical publications about energy resources.*)

6.8 Exercises

E.6.1. What is meant by embodied energy per kilogram of a metal? Why does it differ from the free energy of formation of the oxide, carbonate, or sulfide from the elements that make it up?

E.6.2. What is meant by the process energy per kilogram for casting a metal? Why does it differ from the latent heat of melting of the metal?

E.6.3. Make a bar chart of CO_2 footprint divided by embodied energy using data from the data sheets of Chapter 12. Which material has the highest ratio? Why?

E.6.4. The embodied energies and CO_2 footprints for woods, plywood, and paper do not include a credit for the energy and carbon stored in the wood itself, for the reasons explained in the text. Recalculate these, crediting them with sequestering carbon by subtracting out the stored contribution (take it to be 2.8 kg CO_2 per kg). Is there a net saving?

E.6.5. Rank the three common commodity materials *low carbon steel*, *age-hardening aluminum alloy*, and *polyethylene* by embodied energy/ kg and embodied energy/m^3, using data drawn from the data sheets of Chapter 12 of this book (use the means of the ranges given in the databases). Now rank them by embodied energy per unit stiffness and embodied energy per unit strength. What do you learn?

E.6.6. Iron is made by the reduction of iron oxide, Fe_2O_3, with carbon; aluminum by the electrochemical reduction of Bauxite, basically Al_2O_3. The enthalpy of oxidation of iron to its oxide is 5.5 MJ/kg, that of aluminum to its oxide is 20.5 MJ/kg. Compare these with the embodied energies of cast iron and of aluminum, retrieved from the data sheets of Chapter 12 of this book (use means of the ranges given there). What conclusions do you draw?

E.6.7. Estimate the energy to mold PET by assuming it to be equal to the energy required to heat PET from room temperature to its melting temperature, T_m. Compare this with the actual molding energy. You will find the molding energy, the specific heat, and the melting temperature in the data sheet for PET in Chapter 12 of this book (use means of the ranges). Does the estimate explain the value for molding energy?

E.6.8. The data sheets of Chapter 12 list eco-indicator values where these are available. As explained in the text, eco-indicator values are a normalized, weighted sum involving resource consumption, emissions, and estimates of impact factors. Plot eco-indicator values against embodied energy (a much simpler measure of impact). Is there a correlation?

Exercises using the ces software

E.6.9. Figures 6.8 and 6.9 of the text are plots of the embodied energy of materials per kg and per m^3. Use CES to make similar plots for the carbon footprint. Use the Advanced facility in the axis selection window to make the one for kg CO_2/m^3 by multiplying kg CO_2/kg by the density in kg/m^3.

E.6.10. Plot a bar chart for the embodied energies of metals and compare it with one for polymers, on a "per unit yield strength" basis, using CES. You will need to use the Advanced facility in the axis-selection window to make the function:

$$\textit{Energy per unit strength} = \frac{\textit{Embodied energy} \times \textit{Density}}{\textit{Yield strength}}$$

Which materials are attractive by this measure?

E.6.11. Compare the eco-indicator values of materials with their embodied energy per unit volume. To do so, make a chart with (Embodied energy × Density) on the x axis and Eco-indicator value on the y axis. (Ignore the data for foams since these have an artificially inflated volume.) Is there a correlation between the two? Is it linear? Given that the precision of both could be in error by 10%, are they significantly different measures? Does this give us a way of estimating, approximately, eco-indicator values where none are available?

E.6.12. Plot material price against annual production. Is there a correlation?

Eco-audits and eco-audit tools

7.1 Introduction and synopsis

An *eco-audit* is a fast initial assessment. It identifies the phase of life—material, manufacture, transport, use, disposal—that carries the highest demand for energy or creates the greatest burden of CO_2. It points the finger, so to speak, identifying where the problems lie. Often, one phase of life is, in ecoterms, dominant, accounting for 80% or more of the energy and carbon totals. This difference is so large that the imprecision in the data and the ambiguities in the modeling, discussed in Chapter 3, are not an issue; the dominance remains even when the most extreme values are used. It then makes sense to focus first on this dominant phase, since it is here that the potential gains through innovative material choice are greatest.

Examples of products analyzed in the text.

As we shall see later, material substitution has more complex aspects—there are trade-offs to be considered (see Chapters 8 and 9) —but for now we focus on the simple audit.

The main purpose of an eco-audit is *comparison*, allowing alternative design choices to be explored rapidly. To do this, it is unnecessary to include the last nut and bolt; indeed, with the exception of electronics and precious metals, it is usually enough to account for the few components that make up 95% of the mass of the product, assigning a "proxy" energy and CO_2 to those that are not directly included. The output, of course, is approximate; but if the comparison reveals differences that are large, robust conclusions can be drawn.

This chapter is based around case studies; the exercises at the end propose more. They can be tackled using data from the data sheets in Chapter 12 of this book. Software packages now exist that make the job easier. One is introduced in the appendix.

7.2 Eco-audits

Figure 7.1 shows the procedure for the eco-audit of a product. The *inputs* are of two types. The first are drawn from a user-entered *bill of materials*, *process choice, transport requirements, duty cycle* (the details of the energy and intensity of use), and *disposal route*, shown at the top left. Second, data for embodied energies, process energies, recycle energies, and carbon intensities are drawn from a database of material properties; those for the energy and carbon intensity of transport and the use energy are drawn from lookup tables like those in Chapter 6 (top right of the figure). The *outputs* are the energy or carbon footprint of each phase of life, presented as bar charts and in tabular form. The procedure is best illustrated by a case study of extreme simplicity—that of a PET drink bottle—since this allows the inputs and outputs to be shown in detail. The later case studies are for more complex products, presented in less detail.

The inputs. One brand of bottled water—we will call it *Alpure*—is sold in 1-liter PET bottles with polypropylene caps (see Figure 7.2). One bottle weighs 40 grams; its cap weighs 1 gram. The bottles and caps are molded, filled with water at a source of sparkling purity located in the French Alps, and transported 550 km to London, England, by 14-tonne truck. Once there, the bottles are refrigerated for two days, on average, before appearing on the tables of the restaurant where they are consumed, adding significantly to the diners' bills. The restaurant has an environmental policy: all

Inputs

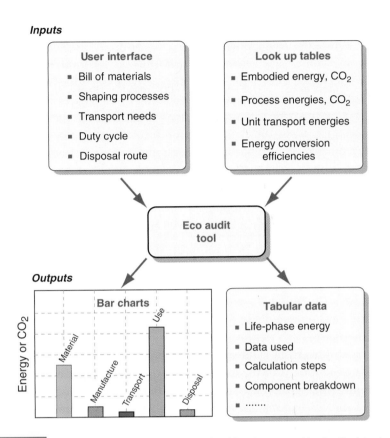

FIGURE 7.1   *The energy audit method. User-defined inputs are combined with data drawn from databases of embodied energy of materials, processing energies (Chapter 12 of this book), transport type (Table 6.7), and energy conversion efficiencies (Tables 6.5 and 6.6) to create the energy breakdown. The same tool can be used for an assessment of the CO_2 footprint.*

Alpure water

FIGURE 7.2   *A 1-litre PET water bottle. The calculation is for 100 units.*

plastic and glass bottles are sent for recycling. We use these data for the case study, taking 100 bottles as the unit of study, requiring $1\,m^3$ of refrigerated space.

The eco-audit procedure has five steps, described here for energy. An audit for CO_2 follows the same steps.

1. *Materials.* A bill of materials is drawn up, listing the mass of each component used in the product and the material of which it is made, as on the left of Table 7.1. Data for the embodied energy (MJ/kg) and CO_2 (kg/kg) per unit mass for each material are

| Table 7.1 | The bill of materials and processes for a PET bottle (100 bottles) | | | | | | | |

Bill of materials				Material	Material	Process	Process
Component	Material	Process	Mass m kg	Energy H_m MJ/kg*	CO_2 kg/kg*	Energy H_p MJ/kg*	CO_2 kg/kg*
Bottle, 100 units	PET	Molded	4	84	2.35	6.8	0.79
Cap, 100 units	PP	Molded	0.1	95	2.7	8.6	0.27
Dead weight (100 liters of water)	Water		100				
		Total mass	104.1				

*From the data sheets of Chapter 12.

retrieved from the database—here, the data sheets of Chapter 12, using the means of the ranges listed there (right side of the table). Multiplying the mass of each component by its embodied energy and summing gives the total material energy—the first bar of the bar chart.

2. *Manufacture.* The audit focuses on primary shaping processes since they are generally the most energy-intensive steps of manufacture. These are listed against each material, as in Table 7.1. The process energies and CO_2 per unit mass are retrieved from the database, as on the right side of the table. Multiplying the mass of each component by its primary shaping energy and summing gives an estimate of the total processing energy, the second bar of the bar chart.

On a first appraisal of the product, it is frequently sufficient to enter data for the components with the greatest mass, accounting for perhaps 95% of the total. The residue is included by adding an entry for "residual components," giving it the mass required to bring the total to 100% and selecting a proxy material and process. "Polycarbonate" and "molding" are good choices because their energies and CO_2 lie in the midrange of those for commodity materials.

3. *Transport.* This step estimates the energy for transportation of the product from manufacturing site to point of sale. For the water bottle, this is dominated by the transport of the filled bottles from the French Alps to London, a distance of 550 km. The energy demands of transport modes were described in Chapter 6 (Table 6.7); that for a 14-tonne truck is 0.9 MJ/tonne.km. Multiplying this figure by the mass of the product and the distance travelled provides the estimate. It is not just the bottles that travel 550 km, it is also the water they contain. This is included in the bill of materials of Table 7.1 to ensure that its mass is included in auditing the transport phase.

4. *The use phase.* The use phase requires a little explanation. There are two different classes of contribution:

   - Some products are (normally) static but require energy to perform their function; electrically powered products such as hairdryers, electric kettles, refrigerators, power tools, and space heaters are examples. Even apparently unpowered products such as household furnishings or unheated buildings still consume some energy in cleaning, lighting, and maintenance. The first class of contribution, then, relates to the power consumed by, or on behalf of, the product itself.

   - The second class is associated with transport. Products that form part of a transport system or are carried around in one add to its mass and thereby augment its energy consumption and CO_2 burden. The transportation table, Table 6.7, lists the energy and CO_2 penalty per unit weight and distance. Multiplying this figure by the product weight and the distance over which it is carried gives an estimate of the associated use-phase energy and CO_2.

All energies are related back to primary energy, meaning oil, via oil-equivalent factors for energy conversion discussed earlier (Tables 6.5 and 6.6). Retrieving these and multiplying by the power and the duty cycle—the usage over the product life—gives an estimate of the oil-equivalent energy of use.

The PET bottle is a static product. Energy is consumed on its behalf via refrigeration for two days. The energy requirements for refrigeration, based on A-rated appliances, are 10.5 MJ/m^3 per day for refrigeration at 4°C and 13.5 MJ/m^3 per day for freezing at −5°C, using electrical power in both cases. The use energy is chosen to give the value for refrigeration.

5. *Disposal.* As explained in Chapter 4 there are five options for disposal at the end of life: *landfill, combustion for energy recovery, recycling, re-engineer* and *reuse* (Figure 4.2). A product at end of first life has the ability to return part or all of its embodied energy. This at first sounds wrong – much of the "embodied" energy was not embodied at all but was lost in the inefficiencies of the processing plant, and even when it is still there, it is, for metals and ceramics, inaccessible: the only easy way to recover energy directly is by combustion, not an option for steel, concrete or brick. But think of it another way. If the materials of the product are recycled or the product itself is re-engineered or reused, a need is filled without drawing on virgin material, giving an energy credit. Carbon release works in the same way, with one little twist: one end-of-life option, combustion, recovers some energy but in doing so it releases CO_2.

Table 7.2 lists the path for each option and first-order estimates for the energies involved to allow their approximate evaluation in the case studies and exercises that follow. The energy cost of transport to landfill is negligible compared with the other energies associated with life. Recycling recovers the difference between the original embodied energy and the energy of recycling. Re-engineering and reuse recover (by filling a need without using new material) almost all the original embodied energy. The data sheets of Chapter 12 provide estimates for recycle energy H_{rc} as well as for the original embodied energy H_m. Where no data appear for recycling it is because the material cannot be recycled. The carbon credit is treated in a similar

Table 7.2 Disposal route and energy balance	
Disposal route	**First-order estimate for energy**
1. *Landfill.* Collect and transport to landfill site.	Negligible.
2. *Combust for heat recovery.* Collect, sort combustibles, combust.	Recover calorific value.
3. *Recycle.* Collect, sort by material family and class, recycle.	Recover difference between embodied energy H_m and recycle energy H_{rc}.
4. *Re-engineer.* Collect, dismantle, replace or upgrade components, re-assemble.	Recover most of embodied energy H_m. Use 0.9 H_m as estimate.
5. *Reuse.* Market as "pre-owned" product via trading outlets, websites etc.	Recover embodied energy H_m.

Table 7.3	Recycle energy and CO_2 for PET					
Component	Material	Mass m kg	Recycle energy H_{rc} MJ/kg*	Recycle CO_2 kg/kg*	$m.H_{tot}$ MJ	$m.(CO_2)_{tot}$ kg
Bottle, 100 units	PET	4	35	0.98	−188	−5.6

*From the data sheets of Chapter 12.

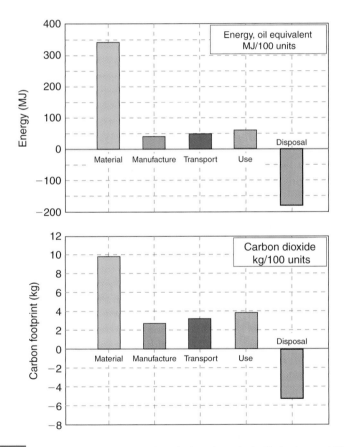

FIGURE 7.3 *The energy and the carbon footprint bar charts for bottled water per 100 units.*

way. Recovered energy and carbon credit appears as negative on the eco-audit bar-charts.

The outputs. Figure 7.3 and Table 7.4 show the outputs. What do we learn? The largest contribution to energy consumption and CO_2 generation

Table 7.4	PET bottle, energy, and CO_2 summary, 100 units			
Life phase	**Life energy (MJ)**	**Energy (%)**	**Life CO_2 (kg)**	**CO_2 (%)**
Material	344	68	9.6	48
Manufacture	36	7	3.2	16
Transport	48	10	3.4	17
Use	74	15	3.7	19
Total	**502**	**100**	**19.9**	**100**

derives from the production of the polymers used to make the bottle. The second largest is that of the short, two-day, refrigeration. The seemingly extravagant part of the life cycle—that of transporting water, 1 kg per bottle, 550 km from the French Alps to the diners' tables in London—contributes 10% of the total energy and 17% of the total carbon. If genuine concern is felt about the eco-impact of drinking Alpure water, then (short of giving it up) it is the bottle that is the primary target. Could it be made thinner, using less PET? (Such bottles are 30% lighter today than they were 15 years ago.) Is there a polymer that is less energy intensive than PET? Could the bottles be made reusable and of sufficiently attractive design that people would want to reuse them? Could recycling of the bottles be made easier? These are design questions, the focus of the lower part of Figure 3.11. Methods for approaching them are detailed in Chapters 8 and 9.

An overall reassessment of the eco-impact of the bottles should, of course, explore ways of reducing energy and carbon in all phases of life, not just one, seeking the most efficient molding methods, the least energy-intensive transportation mode (32-tonne truck, barge) and—an obvious step—minimizing the refrigeration time.

7.3 Case study: an electric kettle

Figure 7.4 shows a typical 2 kW electric kettle. The kettle is manufactured in Southeast Asia and transported to Europe by air freight, a distance of 12,000 km. Table 7.5 lists the materials. The kettle boils 1 liter of water in 3 minutes. It

FIGURE 7.4 *A 2 kW jug kettle.*

Table 7.5	Electric kettle bill of materials, life: three years					
Component	**Material**	**Process**	**Mass, kg**	**Material energy, H_m MJ/kg***	**Process energy, H_p MJ/kg***
Kettle body	Polypropylene	Polymer molding	0.86	94	8.6
Heating element	Ni-Cr alloy	Def. processing	0.026	130	2.6
Casing, heating element	Stainless steel	Def. processing	0.09	81	3.4
Thermostat	Nickel alloys	Def. processing	0.02	72	2.1
Internal insulation	Alumina	Power forming	0.03	52	27
Cable sheath, 1 meter	Natural rubber	Polymer molding	0.06	66	7.6
Cable core, 1 meter	Copper	Def. processing	0.015	71	2.0
Plug body	Phenolic	Polymer molding	0.037	90	13
Plug pins	Brass	Def. processing	0.03	72	2.3
Packaging, padding	Polymer foam	Polymer molding	0.015	110	11
Packaging, box	Cardboard	Construction	0.13	28	0.5
Other small components	*Proxy material:* polycarbonate	*Proxy process:* polymer molding	0.04	110	11
		Total mass	**1.3**		

* From the data sheets of Chapter 12.

is used to do this, on average, twice per day, 300 days per year, over a life of three years. At end of life, the kettle is sent to landfill. How is energy consumption distributed across the phases of the kettle's life?

Figure 7.5 shows the energy breakdown. The first two bars—materials (120 MJ) and manufacture (10 MJ)—are calculated from the data in the table by multiplying the embodied energy by the mass of each component and then summing. Air freight consumes 8.3 MJ/tonne.km (Table 6.7), giving 129 MJ/kettle for the 12,000 km transport. The duty cycle (6 minutes per day, 300 days a year for three years) at full power consumes 180 kW.hr of electrical power. The corresponding consumption of fossil fuel and emission of CO_2 depends on the energy mix and conversion efficiency of the host country (Table 6.6). If this were Australia, for example, the factors from Table 6.6 give, for the use phase, 1800 MJ oil equivalent and 128 kg CO_2, as shown in the figure. At the end of life the kettle is dumped, at an energy cost of 0.2 MJ.

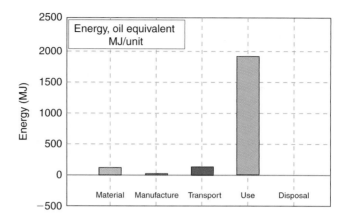

FIGURE 7.5 *The energy bar chart for the electric kettle.*

The use phase of life consumes far more energy than all the others put together. Despite using it for only 6 minutes per day, the electric power (or, rather, the oil equivalent of the electric power) accounts for 88% of the total. Improving ecoperformance here has to focus on this use energy— even a large change, 50% reduction, say, in any of the other uses makes insignificant difference. Heat is lost through the kettle wall. Selecting a polymer with lower thermal conductivity or using an insulated double wall could help here; it would increase the embodied energy of the material bar, but even doubling this leaves it small. A full vacuum insulation would be the ultimate answer; the water not used when the kettle is boiled would then remain hot for long enough to be useful the next time it is needed. The seeming extravagance of air-freight shipping accounts for only 6% of the total energy. Using sea freight instead increases the distance to 17,000 km but reduces the transport energy per kettle to a mere 0.2% of the total.

This dominance of the use phase of energy (and of CO_2 emission) is characteristic of small electrically powered appliances. Further examples can be found in the next case study and the exercises at the end of this chapter.

7.4 Case study: a coffee maker

The 640 Watt coffee maker shown in Figure 7.6 makes four cups of coffee in 5 minutes (requiring full power) and then keeps the coffee hot for a subsequent 30 minutes, consuming one sixth of full power. The housing is injection-molded polypropylene, the jug is glass, there are a number of small steel and aluminum parts, a heating element, and, of course, a cable and plug. The control system has some simple electronics and an LED indicator. Each brew requires a filter paper that is subsequently discarded. We take the life of the coffee maker to be five years, over which time it is used once per day. Table 7.6 summarizes the bill of materials. At the end of life the product is returned to the maker, where it is disassembled and the polypropylene parts are separated and recycled.

Figure 7.7 shows the breakdown, calculated as in the previous case study. Here electronics (Table 6.4) are included, assigning them an energy of 3000 MJ/kg. The first three bars—materials (170 MJ), manufacture (14 MJ), and transportation (5 MJ)—are all small compared to the energy of use. One use cycle uses the equivalent of 10 minutes of full 640 W power. Over five years this becomes 194 kW.hr of electrical power, which, if generated in the United States (Table 6.6), corresponds to an equivalent oil consumption of 1380 MJ and CO_2 emission of 105 kg. Each use also consumes one filter paper—1,825 of them over life—each weighing 2 grams. The embodied energy of paper, from the data

A coffee maker.

sheet in Chapter 12, is 28 MJ/kg, so this represents an additional 100 MJ. At the end of life the polypropylene is recycled, requiring 40 MJ/kg to do so, thereby saving the difference between this and the embodied energy of 94 MJ/kg. (54 MJ/kg)—a return of 49 MJ per unit.

There is nothing that can be done to recover the electrical power once it is used, but it *is* possible to reduce it by replacing the glass jug with a stainless steel vacuum jug, thereby eliminating the need for a heater to keep the coffee warm. The embodied energy of stainless steel is three times greater than that of glass, so it is necessary to check that this redesign really does save energy over the product's life—a task left to the exercises at the end of this chapter.

7.5 Case study: a portable space heater

The space heater shown in Figure 7.8 is carried as part of the equipment of a light goods vehicle used for railway repair work. It burns 0.66 kg of liquid propane gas (LPG) per hour, delivering an output of 9.3 kW (32,000 BTU). The air flow is driven by a 38 W electric fan. The heater weighs 7 kg. The (approximate) bill of materials is listed in Table 7.7. The product is manufactured in India and shipped to the United States by sea freight (15,000 km), then carried by 32-tonne truck for a further 600 km

Table 7.6	Coffee maker142 bill of materials, life: five years					
Component	Material	Process	Mass, kg		Material energy, MJ/kg*	Process energy, MJ/kg*
Housing	Polypropylene	Polymer molding	0.91		94	8.6
Small steel parts	Steel	Def. processing	0.12		81	3.4
Small aluminum parts	Aluminum	Def. processing	0.08		210	2.6
Glass jug	Glass (Pyrex)	Molded	0.33		25	8.2
Heating element	Ni-Cr alloy	Def. processing	0.026		130	2.6
Electronics and LED	Electronics	Assembled	0.007		3000	130
Cable sheath, 1 meter	PVC	Polymer extrusion	0.12		66	7.6
Cable core, 1 meter	Copper	Def. processing	0.035		71	2.0
Plug body	Phenolic	Polymer molding	0.037		90	13
Plug pins	Brass	Def. processing	0.03		72	2.3
Packaging, padding	Polymer foam	Polymer molding	0.015		110	11
Packaging, box	Cardboard	Construction	0.125		28	0.5
Other components	*Proxy material:* Polycarbonate	*Proxy process:* Polymer molding	0.04		110	11
		Total mass	**1.9**			

*From the data sheets of Chapter 12 and Table 6.4.

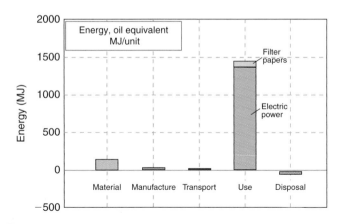

FIGURE 7.7 *The energy breakdown for the coffee maker.*

FIGURE 7.8   *A space heater powered by liquid propane gas (LPG).*

Table 7.7   LPG space heater bill of principal materials, life: three years

Component	Material	Process	Mass (kg)	Mat. CO_2 kg/kg*	Proc. CO_2 kg/kg*
Heater casing	Low C steel	Def. processing	5.4	2.5	0.19
Fan	Low C steel	Def. processing	0.25	2.5	0.19
Heat shield	Stainless steel	Def. processing	0.4	5.1	0.27
Motor, rotor and stator	Iron	Def. processing	0.13	1.0	0.21
Motor, conductors	Copper	Def. processing	0.08	5.7	0.17
Motor, insulation	Polyethylene	Polymer extrusion	0.08	2.1	0.51
Connecting hose, 2 meters	Natural rubber	Polymer molding	0.35	1.5	0.61
Hose connector	Brass	Def. processing	0.09	6.3	0.18
Other components	*Proxy material:* polycarbonate	*Proxy process:* polymer molding	0.22	5.6	0.86
		Total mass	**7.0**		

*From the data sheets of Chapter 12.

to the point of sale. It is anticipated that the vehicle carrying the heater will travel, on average, 420 km per week, over a three-year life, and that the heater itself will be used for 2 hours per day for 10 days per year. At end of product life, the carbon steel components are recycled.

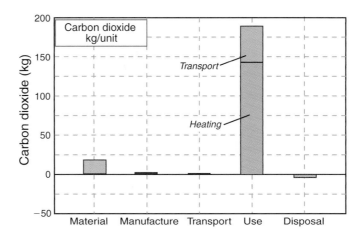

FIGURE 7.9   *The energy breakdown for the space heater. The use phase dominates.*

This is a product that uses energy during its life in two distinct ways. First there is the electricity and LPG required to make it function. Second there is the energy penalty that arises because it increases the weight of the vehicle that carries it by 7 kg. How much CO_2 does the product release over its life? And which phase of life releases the most?

Figure 7.9 shows the CO_2 emission profile. The first two bars—materials (19 kg) and manufacture (1.7 kg)—are calculated from the data in the table. Transport releases only 3.5 kg per unit. The power consumed by burning LPG for heat (9.3 kW) far outweighs that used to drive the small electric fan motor (38 W), so it is the CO_2 released by burning LPG that we evaluate here. It is less obvious how the static use for generating heat, drawn for only 20 hours per year, compares with the extra fuel consumed by the vehicle because of the product weight—remembering that, as part of the equipment, it is lugged over 22,000 km per year. The figure shows that the CO_2 of use outweighs all other contributions, here accounting for 90% of the total, as it does with most energy-using products. Of this, 34% derives from burning gas and 76% from the additional fuel consumed by carrying the heater to the sites where it is used. The CO_2 burden to recycle steel, from the data sheet in Chapter 12, is about 0.7 kg/kg, saving the difference between this and that for primary production (2.5 kg/kg)—a net saving of 10 kg.

7.6 Auto bumpers: exploring substitution

The bumpers of a car are heavy; reducing their weight can save fuel. Here we explore the replacement of a low alloy steel bumper with one of equal

Table 7.8	The analysis of the material substitution				
Material of bumper	Mass kg	Material energy, MJ/kg*	Material energy, MJ	Use energy, MJ	Total: material plus use, MJ
Low alloy steel	14	35	490	7400	7890
Age-hardened Alu	10	210	2100	5300	7400

*From the data sheets of Chapter 12.

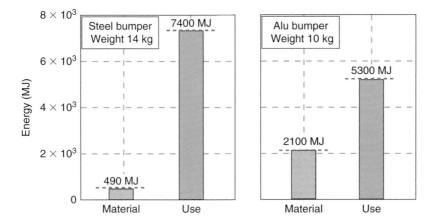

FIGURE 7.10 *The comparison of the energy audits of a steel and an aluminum fender for a family car.*

performance made from an age-hardening aluminum alloy, on a gasoline-powered family car. The steel bumper weighs 14 kg; the aluminum substitute weighs 10 kg, a reduction in weight of 28%. But the embodied energy of aluminum is much higher than that of steel. Is there a net saving? As in the previous case study, we take a vehicle life of 10 years and a car driven 25,000 km per year. Table 7.8 lists the energies involved. We take the energy penalty of weight for a gasoline-powered car (Table 6.7) to be 2.06 MJ/tonne.km.

The bar charts of Figure 7.10 compare the material and use energy, assuming the use of virgin material for both the steel and the aluminum bumper. The substitution of steel by aluminum results in a large increase in material energy but a drop in use energy. The right-side columns of the table list values and the totals; the aluminum substitute wins (it has a lower total) but not by much—the breakeven comes at about 200,000 km. And it costs more.

But this is not quite fair. A product like this would, if possible, incorporate recycled as well as virgin material. It is possible, using the data given in the data sheets of Chapter 12, to correct the material energies for the recycle content. This is left to the appendix, where we discuss eco-audit tools.

7.7 Family car: comparing material energy with use energy

Here we move up in scale. Argonne National Laboratory, working with the U.S. Department of Energy, has developed a model (GREET) to evaluate energy and emissions associated with vehicle life. Table 7.9 lists the

Table 7.9 Material content of a conventional family car and one made of lightweight materials

Material	Conventional ICE vehicle, kg	Lightweight ICE vehicle, kg	Material energy H_m MJ/kg*
Carbon steel	*839*	*254*	32
Stainless steel	0.0	5.8	81
Cast iron	*151*	*31*	17
Wrought aluminum (10% recycle content)	*30*	*53*	200
Cast aluminum (35% recycle content)	*64*	*118*	149
Copper/brass	26	45	72
Magnesium	0.3	3.3	380
Glass	39	33	15
Thermoplastic polymers (PU, PVC)	94	65	80
Thermosetting polymers (polyester)	55	41	88
Rubber	33	17	110
CFRP	*0.0*	*134*	273
GFRP	0.0	20	110
Platinum, catalyst (*Table 6.3*)	0.007	0.003	117000
Electronics, emission control etc (*Table 6.4*)	0.27	0.167	3000
Other (*proxy material: polycarbonate*)	26	18	110
Total mass	*1361*	*836*	

*From the data sheets of Chapter 12 and Tables 6.3 and 6.4.

bill of materials for two of the vehicles they analyze: a conventional mid-sized family car with an internal combustion engine (ICE) and a vehicle of similar size made of lightweight materials, with the biggest differences italicized in bold. The total mass is shown at the bottom of the columns. "Lightweighting" reduces it by 39%.

The data sheets of Chapter 12 provide the embodied energies of the materials. Fuel consumption scales with weight in ways that are analyzed in Chapter 9; for now we use the results that a conventional car of this weight consumes 3.15 MJ/km; the lighter one consumes 2.0 MJ/km.[1] There is enough information here to allow an approximate comparison of embodied energy and the use of the two vehicles, assuming both are driven 25,000 km per year for 10 years.

The bar charts of Figure 7.11 show the comparison. The input data are of the most approximate nature, but it would take very large discrepancies to change the conclusion: the energy consumed in the use phase of both vehicles greatly exceeds that embodied in their materials. The use of lightweight materials increases the embodied energy by 43% but reduces the much larger fuel-energy consumption by 37%. The result is a net gain: the sum of the material and use energies for the lightweight vehicle is 30% less than that of the conventional one.

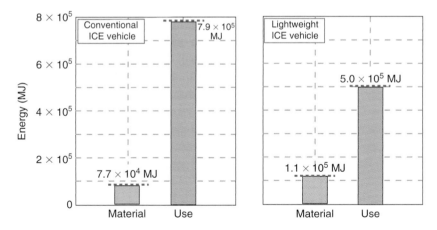

FIGURE 7.11 *The comparison of the energy audits of a the conventional and the lightweight family car.*

---

[1] 1 MJ/km = 2.86 liters/100 km = 95.5 miles per U.K. gallon = 79.5 miles per U.S. gallon.

7.8 Energy flows and payback time of a wind turbine

Now we move up in scale a second time. Wind energy (see the Chapter 10 title page figure) is attractive in a number of ways. It is renewable, it is not dependent on fuel supplies from diminishing resources sited in countries other than your own, it does not pose a threat in the hands of hostile nations, and, by its nature, it is distributed and thus difficult to disrupt. But is it energy efficient? Energy has to be invested to build the turbine; how long does it take for the turbine to pay back on the investment?

Table 7.10 lists the bill of principal materials for a 2 MW land-based turbine. The information is drawn in part from a study conducted for Vestas Wind Systems,[2] in part from the Technical Specification of Nordex

Table 7.10	Approximate bill of materials for onshore wind turbine			
Subsystem	Component	Material	Process	Mass (kg)
Tower (165 tonnes)	Structure	Low carbon steel	Def. processing	164,000
	Cathodic protection	Zinc	Casting	203
	Gears	Stainless steel	Def. processing	19,000
	Generator, core	Iron (low C steel)	Def. processing	9000
	Generator, conductors	Copper	Def. processing	1000
Nacelle (61 tonnes)	Transformer, core	Iron	Polymer molding	6000
	Transformer, conductors	Copper	Def. processing	2000
	Transformer, conductors	Aluminum	Def. processing	1700
	Cover	GFRP	Composite forming	4000
	Main shaft	Cast iron	Casting	12,000
	Other forged components	Stainless steel	Def. processing	3000
	Other cast components	Cast iron	Casting	4000
Rotor (34 tonnes)	Blades	CFRP	Composite forming	24,500
	Iron components	Cast iron	Casting	2000
	Spinner	GFRP	Composite forming	3000
	Spinner	Cast iron	Casting	2200
Foundations (832 tonnes)	Pile and platform	Concrete	Construction	805,000
	Steel	Low carbon steel	Def. processing	27,000
Transmission	Conductors	Copper	Def. processing	254
	Conductors	Aluminum	Def. processing	72
	Insulation	Polyethylene	Polymer extrusion	1380
			Total mass	1,100,000

[2]Elsam Engineering A/S, Life-cycle assessment of offshore and onshore sited wind farms, October 2004. This document lists the quantities of the more significant materials and the weight of each subsystem. The nacelle itself is made up of a number of smaller parts, some of which are difficult to assess due to the limited information contained in the report.

Energy,[3] and in part from Vestas' own report[4] scaling its data according to weight. Some energy is consumed during the life of the turbine (design life: 25 years), mostly in the transport associated with maintenance. This was estimated from information on inspection and service visits from the Vestas report, together with the estimated distance traveled.

The net energy demands of each phase of life are summarized in Table 7.10 (in units of MJ in the second column and in kW.hr in the third). The turbine is rated at 2 MW, but it produces this power only when the wind conditions are right. This is measured by the *capacity factor*—the fraction of peak power delivered, on average, over a year. We take this to be 25%, giving an annual energy output of 4.4×10^6 kW.hr/year.[5] The *energy payback time* is then the ratio of the total energy invested in the turbine (including maintenance) and the expected average yearly energy production:

$$\text{Payback time} = \frac{5.3 \times 10^6 \text{ kW.hr}}{4.4 \times 10^6 \text{ kW.hr/}yr} = 1.2 \text{ years} = 14.5 \text{ months}$$

The total energy generated by the turbine over a 25-year life is about 1×10^8 kW.hr, roughly 20 times that required to build and service it.

The Vestas LCA study for this turbine, a much more detailed analysis of which only some of the inputs are published, arrives at a payback time of eight months using a capacity factor of 50% (correcting to 25% gives 16 months). A recent study at the University of Wisconsin-Madison[6] finds that wind farms have a high "energy payback" (ratio of energy produced compared to energy expended in construction and operation), larger than that of either coal or nuclear power generation. In the study, three Midwestern wind farms were found to generate between 17 and 39 times more energy than is required for their construction and operation, whereas coal-fired power stations generate, on average, 11 times as much and nuclear plants 16 times as much. And, of course, coal-fired power stations emit CO_2.

The construction of the wind turbine itself carries a carbon footprint. Using data CO_2 from the data sheets of Chapter 12 and Table 6.7 gives the values shown in the last column of Table 7.11—a total output of 1,400

---

[3]Nordex N90 Technical Description, Nordex Energy GmbH, April 2004.

[4]Vestas, Life-cycle assessment of offshore and onshore sited wind turbines, Vestas Wind Systems A/S, 2005, www.vestas.com.

[5]A study of Danish wind turbines located in particularly favorable sites found a capacity factor of 54%, but this is unusual. A value of 25% is more realistic.

[6]Wind Energy Weekly, Vol. 18, Number 851, June 1999.

Table 7.11	The energy analysis for the construction and maintenance of the wind turbine		
Phase	**Invested energy (MJ)**	**Invested energy (kW.hr)**	**CO_2 emissions (kg)**
Material	1.8×10^7	5.0×10^6	1.3×10^6
Manufacture	1.2×10^6	3.3×10^5	9.7×10^4
Transport	2.8×10^5	7.8×10^4	2×10^4
Use (maintenance)	1.9×10^5	5.3×10^4	1.4×10^4
Total	**1.9×10^7**	**5.3×10^6**	**1.4×10^6**

tonnes of CO_2. But the energy produced by the turbine is almost carbon-free. The life output of 1×10^8 kW.hr, if generated from fossil fuels, would have emitted 21,000 tonnes of CO_2. Thus wind turbines offer power with a much reduced carbon footprint.

The problem with wind power is not energy payback, but the small power output per unit. Even with an optimistic capacity factor of 50%, about 1000 2MW wind turbines are needed to replace the power output of just one conventional coal-fired power station.

7.9 Computer-aided eco-auditing

An eco-audit guides decision making during design or redesign of a product and it points to the aspects of design that a fuller LCA should examine. Early in the design process, the detailed information required for a rigorous LCA is not available, and even if it were, the formal LCA methods are sufficiently burdensome to inhibit their repeated use for rapid "What if?" exploration of alternatives. As we have seen, an eco-audit, though approximate in every way, frequently reveals differences that are sufficiently large to be significant. LCA tools such as SimaPro, GaBi, MEEUP, and the Boustead Model (see the appendix to Chapter 3) can be used in this way, but they were not designed for it. Others are much simpler to use. One—the CES eco-audit tool[7]—is described in the appendix to this chapter. It implements the procedure shown in Figure 7.1, enabling the inputs and delivering the outputs shown there.

---

[7] The CES Eco Audit tool is a standard part of the CES Edu Package developed by Granta Design, Cambridge, U.K., www.GrantaDesign.com.

7.10 Summary and conclusion

Eco-aware product design has many aspects, one of which is the choice of materials. Materials are energy intensive, with high embodied energies and associated carbon footprints. Seeking to use low-energy materials might appear to be one way forward, but this can be misleading. Material choice impacts manufacturing; it influences the weight of the product and its thermal and electrical characteristics and thus the energy it consumes during use, and it influences the potential for recycling or energy recovery at the end of life. It is full-life energy that we seek to minimize.

Doing so requires a two-part strategy, developed in Chapter 3. The first part is an *eco-audit*: a quick, approximate assessment of the distribution of energy demand and carbon emission over a product's life. This provides inputs to guide the second part: that of *material selection to minimize the energy and carbon* over the full life, balancing the influences of the choice over each phase of life—the subject of Chapters 8 and 9. The eco-audit method described here is fast and easy to perform, and although approximate, it delivers information with sufficient precision to enable strategic decision making. This chapter has introduced the procedure and tools—the appendix describes one in more depth—illustrating their use with case studies. The exercises that follow provide opportunities for exploring them further.

7.11 Further reading

Burnham, A., Wang, M. and Wu, Y. (2006), "Development and applications of GREET 2.7", Argonne National Laboratory, ANL/ESD/06-5. www.osti. gov/bridge. *(A report describing the model developed by ANL for the U.S. Department of Energy to analyze life emissions from vehicles).*

Ashby, M. F., Ball, N. and Bream, C. (2008), The CES EduPack eco-audit tool: a white paper, Granta Design, www.GrantaDesign.com. *(A user guide for the CES eco-audit tool.)*

7.12 Appendix: the CES eco-audit tool

The CES Edu software includes, as standard, an eco-audit tool. The double spread of Figure 7.12 is a mock-up of the user interface. It shows the user actions and the consequences. There are five steps, labelled 1 to 5. Actions and inputs are shown on the left in red.

1. *Product definition* allows entry of a descriptive name.

2. *Material, process and end of life* allows entry of the bill of materials and processes: the *material*, the *primary shaping process*, and the

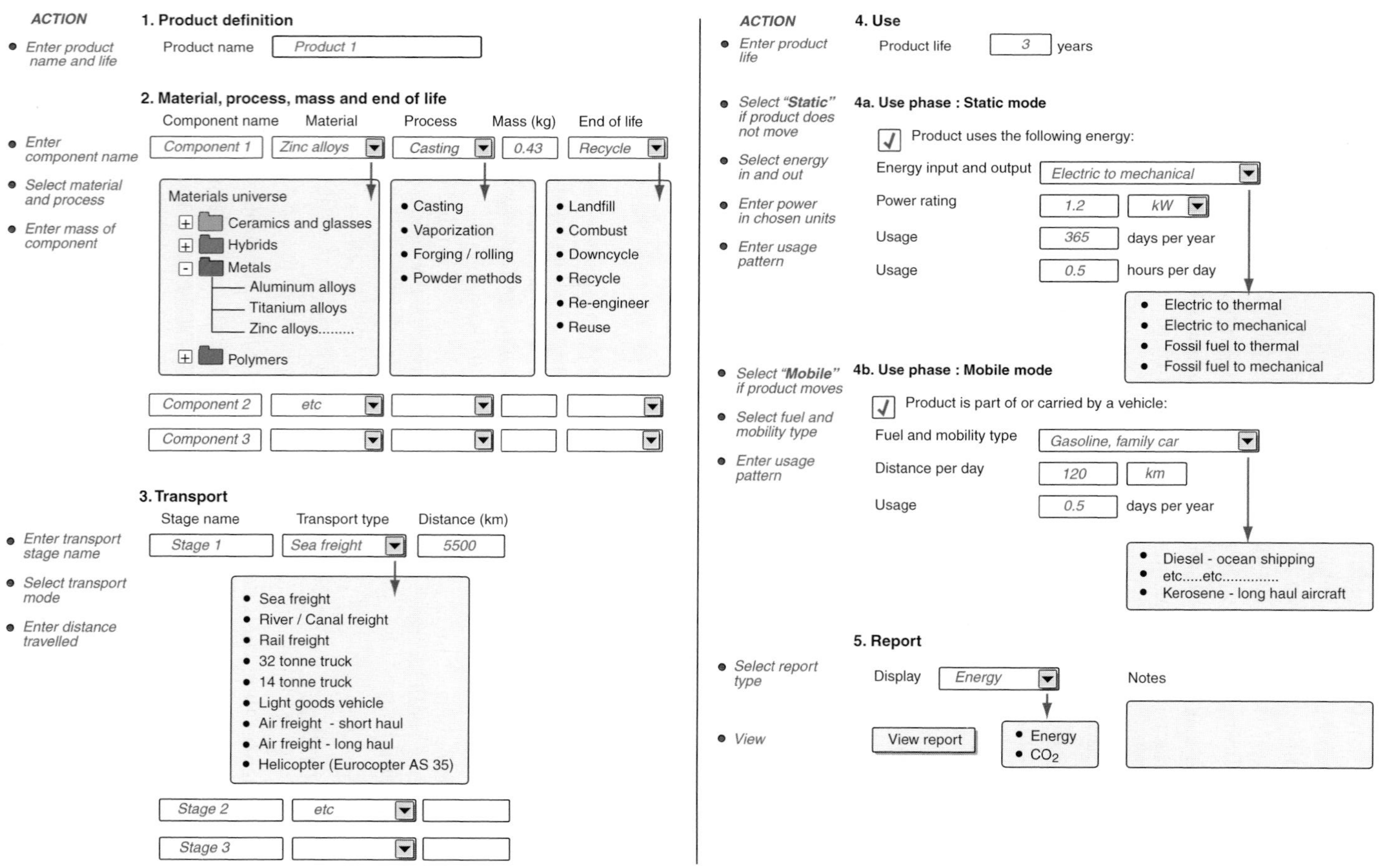

FIGURE 7.12   The CES eco-audit.

mass for each component. The component name is entered in the first box. The material is chosen from the pull-down menu of box 2, opening the CES database of materials attributes.[8] Selecting a material from the tree-like hierarchy of materials causes the tool to retrieve and store its embodied energy and CO_2 footprint per kg. The primary shaping process is chosen from the pull-down menu of box 3, which lists the processes relevant for the chosen material; the tool again retrieves energy and carbon footprint per kg. The next box allows the component weight to be entered in kg. The final box allows choice of disposal route at end of life. On completing a row entry, a new row appears for the next component.

As explained earlier, for a first appraisal of the product, it is frequently sufficient to enter data for the components with the greatest mass, accounting for perhaps 95% of the total. The residue is included by adding an entry for **residual components**, giving it the mass required to bring the total to 100% and selecting a proxy material and process: **polycarbonate** and **molding** are good choices because their energies and CO_2 lie in the midrange of those for commodity materials. Dead weight (weight that is not part of the product but must be transported with it, like the water in the example of this chapter) is entered without choosing a material or process.

The tool multiplies the energy and CO_2 per kg of each component by its mass and totals them. It estimates energy recovered and CO_2 credit as explained in section 7.2. In its present form, the data for materials are comprehensive. The data for processes are rudimentary.

3. *Transport* allows for transportation of the product from manufacturing site to point of sale. The tool allows multistage transport (e.g., shipping followed by delivery by truck). To use it, the stage is given a name, a transport type is selected from the pull-down **Transport type** menu, and a distance is entered in km or miles. The tool retrieves the energy/tonne.km and the CO_2/tonne.km for the chosen transport type from a look-up table like that of Table 6.7 and multiplies them by the product weight and the distance travelled, finally summing the stages.

---

[8] One of the CES Edu Materials databases, depending on which was chosen when CES was opened.

4. *The use phase* allows the two different classes of contribution that were described in Step 4 of Section 7.2: the static mode contribution, 4a, and the mobile contribution, 4b. Clicking the **Static mode** checkbox activates part 4a. Selecting an energy conversion mode causes the tool to retrieve the efficiency (from a look-up table like that of Table 2.1) and to multiply it by the power and the duty cycle, entered here as *use days per year* times *hours per day* times *product life* in years.

As explained earlier, products that are part of a transport system carry an additional energy and CO_2 penalty by contributing to its weight. Clicking the **Mobile mode** checkbox enables part 4b. A pull-down menu allows selection of the type of transport, listed by fuel and mobility type. On entering daily distance, the tool calculates the energy and CO_2 by multiplying product weight and distance carried by the energy or CO_2 per tonne.km drawn from the same look-up table as that for the transport phase.

5. The final step allows the user to select energy or CO_2 as the measure, displaying it as a bar chart and in tabular form. Clicking **Report** then completes the calculation, presenting the results as bar charts like those of Section 7.3 and in tabular form, detailing the inputs.

Case study: a 2000 W electric hairdryer

A hairdryer is a product of which at least one (usually more) can be found in almost every European and U.S. household. Figure 7.13 shows a

FIGURE 7.13 *A 2000W "ionic" diffuser hairdryer.*

contemporary 2000 W dryer, made in Southeast Asia and shipped by sea to Europe, roughly 20,000 km. The bill of materials is reproduced from the tool as Table 7.12. The dryer has an expected life of three years (it is guaranteed for only two), and will be used, on average, for 3 minutes per day for 150 days per year. At end of life the polymeric housing and nozzle subsystem is recycled.

Figure 7.14 and Table 7.13 show part of the output. As with other energy-using products, it is the use phase that dominates. The tool provides more detail: tables listing the material and process energies it retrieved from the database, the energy contribution that can be attributed to each component, and the details of the transport and use-phase calculations. We

Table 7.12	Hairdryer bill of materials and processes			
Subsystem	**Component**	**Material**	**Shaping process**	**Mass (kg)**
Housing and nozzle	Housing	ABS	Polymer molding	0.177
	Inner air duct	Nylons (PA)	Polymer molding	0.081
	Filter	Polypropylene	Polymer molding	0.011
	Diffuser	Polypropylene	Polymer molding	0.084
Fan and motor	Fan	Polypropylene	Polymer molding	0.007
	Casing	Polycarbonate	Polymer molding	0.042
	Motor—iron	Low carbon steel	Def. processing	0.045
	Motor—windings	Copper	Def. processing	0.006
	Motor—magnet	Nickel	Def. processing	0.022
Heater	Heating filament	Nickel-chrome alloys	Def. processing	0.008
	Insulation	Alumina	Ceramic power forming	0.020
	Support	Low carbon steel	Def. processing	0.006
Circuit board and wiring	Board	Phenolics	Polymer molding	0.007
	Conductors	Copper	Def. processing	0.006
	Insulators	Phenolics	Polymer molding	0.012
	Cable sheathing	Polyvinylchloride	Polymer molding	0.005
Cable and Plug	Main cable, core	Copper	Def. processing	0.035
	Cable sleeve	Polyvinylchloride	Polymer molding	0.109
	Plug body	Phenolics	Polymer molding	0.021
	Plug pins	Brass	Def. processing	0.023
Packaging	Rigid foam padding	Rigid polymer Foam	Polymer molding	0.011
	Box	Paper and cardboard	Construction	0.141
Residual components	Residual components	*Proxy material:* polycarbonate	*Proxy process:* polymer molding	0.010
			Total mass	**0.89**

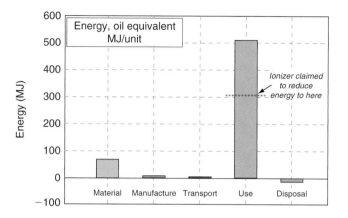

The energy breakdown for the hairdryer, as delivered by the CES eco-audit tool.

Table 7.13	The Energy breakdown for the hairdryer	
Phase of life	**Energy (MJ)**	**Energy (%)**
Material	72.3	11
Manufacture	6.8	1
Transport	88.5	13
Use	506	75
Total	673	100

conclude that improving the heat-transfer efficiency is the most obvious, though not, probably, the easiest way to reduce energy consumption.

It may be possible to reduce the material bar by substitution with less energy-intensive materials or by using a greater recycled content, but the effect of this on the life energy will be small. So the target has to be the heater. In the operation of a hairdryer, most of the heat goes straight past the head. The fraction that is functionally useful is not known, but it is small. Anything that increases this gives energy efficiency. The diffuser (a standard accessory) does just that. The makers of this hairdryer claim another, even more dramatic, development. Incorporated within it is a gas-discharge ionizer. It ionizes the air flowing past it, which, it is speculated, breaks down the water in the hair into smaller droplets, allowing them to evaporate faster.

The manufacturers of ionic hairdryers claim that this process dries the hair twice as fast (it is printed on the box). If true, this claim gives an impressive total life-energy reduction of nearly 40%, as marked on the figure. Although we have found no scientific demonstration for the claim, customer reviews express a market preference for ionic hairdryers.

7.13 Exercises

E.7.1. If the embodied energies and CO_2 used in the Alpure water case study in the chapter are uncertain by a factor of ±25%, do the conclusions change? Mark ±25% margins onto each bar in a copy of Figure 7.3 (you are free to copy it) and then state your case.

E.7.2. Alpure water has proved to be popular. The importers now want to move up-market. To do so they plan to market their water in 1 liter glass bottles of appealing design instead of the rather ordinary PET bottles with which we are familiar from the case study in the chapter. A single 1-liter glass bottle weighs 430 grams, much more than the 40 grams of those made of PET. Critics argue that this marketing-strategy is irresponsible because of the increased weight. The importers respond that glass has lower embodied energy than PET.

Use the methods of this chapter and the data available in Chapter 12 to analyze this situation. What do you conclude?

E.7.3. If the glass container of the coffeemaker of Section 7.4 is replaced by a double-walled stainless steel one weighing twice as much, how much does the total embodied energy of the product change? If this reduces the electric power consumed over the product life by 10%, is the energy balance favorable?

E.7.4. The figure shows a 1700 W steam iron. It weighs 1.3 kg, 98% of which is accounted for by the seven components listed in the table. The iron heats up on full power in 4 minutes and is then used, typically, for 20 minutes on half power. At end of life the iron is dumped as landfill. Create an eco-audit for the iron, assuming that it is used once per week over a life of five years, using data from the data sheets of Chapter 12.

What conclusions can you draw? How might the energy be reduced?

Iron: bill of materials

Component	Mass (kg)	Material	Shaping process
Body	0.15	Polypropylene	Molded
Heating element	0.03	Nichrome	Drawn
Base	0.80	Stainless steel	Cast
Cable sheath, 3 meter	0.18	Polyurethane	Molded
Cable core, 3 meter	0.05	Copper	Drawn
Plug body	0.037	Phenolic	Molded
Plug pins	0.03	Brass	Rolled

E.7.5. The picture shows a 970 W toaster. It weighs 1.2 kg including 0.75 m of cable and plug. It takes 2 minutes, 15 seconds, to toast a pair of slices. It is used to toast, on average, eight slices per day, so it draws its full electrical power for 9 minutes (540 seconds) per day over its design life of three years. The toasters are made locally; transport energy and CO_2 are negligible. At end of life it is dumped. Create an eco-audit for the toaster using data from the data sheets of Chapter 12.

Toaster: bill of materials

Component	Material	Shaping Process	Mass (kg)
Body	Polypropylene	Molded	0.24
Heating element	Nichrome	Drawn	0.03
Inner frame	Low carbon steel	Rolled	0.93
Cable sheath, 0.75 meter	Polyurethane	Molded	0.045
Cable core, 0.75 meter	Copper	Drawn	0.011
Plug body	Phenolic	Molded	0.037
Plug pins	Brass	Rolled	0.03

E.7.6. It is proposed to replace the steel bumper of Case Study 7.6 by one made of CFRP. It is anticipated that the CFRP will weigh 7 kg. Following the procedure of the text, drawing data from the data sheets of Chapter 12, estimate whether, over the life pattern used in the text, there is a net energy saving.

E.7.7. The production of a small car (mass 1000 kg) requires materials with a total embodied energy of 70 GJ, and a further 15 GJ for the manufacturing phase. The car is manufactured in Germany and delivered to the U.S. showroom by sea freight (distance 10,000 km) followed by delivery by heavy truck over a further 250 km (Table 6.7 of the text gives the energy per tonne.km for both). The car has a useful life of 10 years and will be driven, on average, 25,000 km per year, consuming 2 MJ/km. Assume that recycling at end of life recovers 25 GJ per vehicle.

Make an energy-audit bar chart for the car with bars for material, manufacture, distribution, and use. Which phase of life consumes most energy? The inherent uncertainty of current data for embodied and processing energies is considerable; if both were in error by a factor of 2 either way, can you still draw firm conclusions from the data? If so, what steps would do most to reduce life-energy requirements?

E.7.8. The following table lists one European automaker's summary of the material content of a midsized family car. Material proxies for the vague material descriptions are given in brackets and italicized. The vehicle is gasoline-powered and weighs 1800 kg. The data sheets of Chapter 12 provide the embodied energies of the materials; mean values are listed in the table. Table 6.7 gives the energy of use: 2.1 MJ/tonne. km, equating to 3.8 MJ/km for a car of this weight. Use this information to allow a rough comparison of embodied and use energies of the car, assuming it is driven 25,000 km per year for 10 years.

Material content of a family car, total weight 1800 kg

Material	Mass (kg)	Material energy, MJ/kg*
Steel (low alloy steel)	950	32
Aluminum (cast aluminum alloy)	438	220

Material	Mass (kg)	Material energy, MJ/kg*
Thermoplastic polymers (PU, PVC)	148	80
Thermosetting polymers (polyester)	93	88
Elastomers (butyl rubber)	40	110
Glass (borosilicate glass)	40	15
Other metals (copper)	61	72
Textiles (polyester)	47	47
Total mass	1800	

From the data sheets of Chapter 12.

E.7.9. Carry out an energy eco-audit for a product of your choosing. Pick something simple (a polypropylene washing-up bowl, for example) or, more ambitiously, something more complex. Dismantle it, weigh the parts, and use your best judgment to decide what they are made of and how they were shaped.

E.7.10. Conduct a CO_2 eco-audit for the patio heater shown here. It is manufactured in Southeast Asia and shipped 8,000 km to the United States, where it is sold and used. It weighs 24 kg, of which 17 kg is rolled

stainless steel, 6 kg is rolled carbon steel, 0.6 kg is cast brass, and 0.4 kg is unidentified injection-molded plastic (so use a proxy of your own choosing for this). In use it delivers 14 kW of heat ("enough to keep eight people warm"), consuming 0.9 kg of propane gas (LPG) per hour, releasing 0.059 kg of CO_2/MJ. The heater is used for 3 hours per day for 30 days per year, over five years, by which time the owner tires of it and takes it to the recycling depot (only 6 miles/ 10 km away, so neglect the transport CO_2), where the stainless steel is dismembered, and the steel and brass are sent for recycling.

Use data from the text and data sheets to construct a bar chart.

Exercises using the CES eco-audit tool

E.7.11. Repeat the analysis of the PET bottle for the Alpure water case, the first case study in the chapter, entering data from Table 7.1 using the CES eco-audit tool. Then repeat, replacing the PET bottle by a glass one, using the additional information in Exercise E.7.2.

E.7.12. Carry out the eco-audit for the PP electric kettle of Section 7.3, now using the fact that the kettle body and the cardboard packaging are recycled at end of life.

E.7.13. Carry out the eco-audit for the coffeemaker of Section 7.4, now using the fact that it is made entirely out of recycled materials (tick the box "100% recycled").

E.7.14. Carry out the eco-audit for the portable space heater using the data from Section 7.5, assuming that at end of first life, the fan and heat shield are re-engineered to incorporate them into a new product.

E.7.15. Carry out Exercise E.7.5 (the toaster) using the CES eco-audit tool and the bill of materials listed there. Make bar charts for both energy and CO_2. Then tick the box "standard grade" in the calculation, and repeat. Compare the result with using virgin material for making the toaster. "Standard grade" means that the materials contain the typical recycle fraction listed in the data sheets of Chapter 12.

Selection strategies

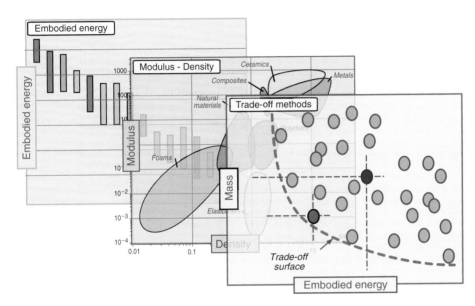

Trade-off surface

8.1 Introduction and synopsis

Life is full of decisions. Which shoes to buy? Which restaurant to eat at? Which camera? Which bike? Which car? Which university? Most of us evolve strategies for reaching decisions, some involving emotional response ("It's cool/I just couldn't resist the color, and besides, Joe/Joanna has one"), others based on cold logic.

It is cold logic that we want here. The strategy then takes the following form:

- Assemble data for the characteristics of the thing you want to select; make a *database*, mental or physical.

- Formulate the characteristics that the thing must have to satisfy your requirement; list the *constraints*.

Three selection tools: bar-charts, bubble charts, and trade-off plots.

- Decide on the ranking criterion you will use to decide which, of the candidate things that meet all the constraints, is the best; choose and apply the *objective*.
- Research the top-ranked candidates more fully to satisfy yourself that nothing has been overlooked; seek *documentation*.

This chapter is about selection using this strategy. It is simpler to start with a product than with a material; the ideas are the same, but the material has added complications. So we start with cars, basing the discussion around the selection of a car to meet given constraints and with two objectives, one of them that of minimizing carbon footprint. Selecting materials uses the same reasoning and method. The rest of the chapter describes how to do it.

8.2 The selection strategy: choosing a car

You need a new car. To meet your needs it must be a midsized four-door family car with a petrol engine delivering at least 150 horsepower—enough to tow your sailboat. Given all of these requirements, you want the car to cost as little to own and run as possible (see Figure 8.1, left side). There are three *constraints* here, but they are not all of the same type:

- The requirements of *four-door family car* and *petrol power* are *simple constraints*—a car *must* have these to be a candidate.

- The requirement of *at least 150 hp* places a lower limit but no upper one on power; it is a *limit constraint*—any car with 150 hp or more is acceptable.

The wish for minimum cost of ownership is an *objective*, a criterion of excellence. The most desirable cars from among those that meet the constraints are those that minimize this objective.

To proceed, you need information about available cars (Figure 8.1, right side). Carmakers' Websites, dealers, car magazines, and advertisements in the national press list such information, including car type and size, number of doors, fuel type, engine power, and price. Car magazines go further and estimate the cost of ownership (meaning the sum of running costs, tax, insurance, servicing and depreciation), listing it as $/mile or €/km.

Now it's decision time (Figure 8.1, central box). The selection engine (you, in this example) uses the constraints to *screen out*, from all the available cars, those that are not four-door gasoline-powered family vehicles with 150 hp. Many cars meet these constraints; the list is still long. You need a way to order it so that the best choices are at the top. That is what the objective is for: it allows you to *rank* the surviving candidates by cost of ownership—those with the lowest values are ranked most highly. Rather

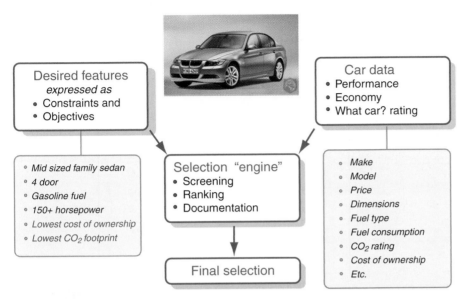

Desired features
expressed as
- Constraints and
- Objectives

Car data
- Performance
- Economy
- What car? rating

- *Mid sized family sedan*
- *4 door*
- *Gasoline fuel*
- *150+ horsepower*
- *Lowest cost of ownership*
- *Lowest CO_2 footprint*

Selection "engine"
- Screening
- Ranking
- Documentation

- *Make*
- *Model*
- *Price*
- *Dimensions*
- *Fuel type*
- *Fuel consumption*
- *CO_2 rating*
- *Cost of ownership*
- *Etc.*

Final selection

FIGURE 8.1 *Selecting a car. The requirements are expressed as constraints and objectives (objectives in blue). Records containing data for cars are screened using the constraints and ranked by the objectives to find the most attractive candidates. These are then explored further by examining documentation.*

than simply choosing the one at the top, it is better to keep the top three or four and seek further *documentation*, exploring their other features in depth (delivery time, size of trunk, comfort of seats, security, and so on) and weighing the small differences in cost against the desirability of these features.

But we have overlooked a second objective, listed in blue on the left of Figure 8.1. You are an environmentally responsible person; you want to minimize the CO_2 rating as well as the cost of ownership. The choice to meet two objectives is more complicated than that to meet just one. The problem is that the car that best satisfies one objective—minimizing cost, for example—might not be the one that minimizes the other (CO_2), and vice versa, so it is not possible to minimize both at the same time. A compromise has to be reached, and that needs *tradeoff methods*.

Figure 8.2 shows such a method. Its axes are the two objectives: cost of ownership and CO_2 rating. Suppose a friend has recommended a particular car, shown as a red dot at the center of the diagram. Your research has revealed cars with combinations of cost and carbon, shown by the other dots. Several—the purple ones in the upper right—have higher values of both; they lie in the "unacceptable" quadrant. Several—the blue ones—either have lower cost or lower carbon, but not both. One—the green one—is both cheaper *and* produces less carbon; it ranks more highly by both objectives. Thus it is the obvious choice.

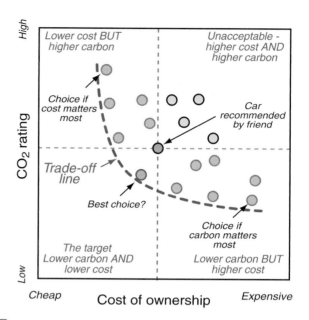

FIGURE 8.2 *The tradeoff between carbon footprint and cost of ownership.*

Or is it? That depends on the value you attach to low carbon footprint. If you think it is a good idea as long as you don't have to pay, the car marked "Choice if cost matters most" is the best. If instead you are ready to pay whatever it takes to minimize CO_2 emissions, the car marked "Choice if carbon matters most" is the one to go for.

All three choices lie, with several others that are compromises between them, on the boundary of the occupied region of the figure. The envelope of these—the broken line—is called the *tradeoff line*. Cars that lie on or near this line have the best compromise combination of cost and carbon. So even if we can't reach a single definitive choice (at least without knowing exactly what you think a low carbon footprint is worth), we have made progress. The viable candidates are those on or close to the tradeoff line. All others are definitely less good.

Methods like this are used as tools for decision making in many fields; in deciding between design options for new products, in optimizing the operating methods for a new plant, in guiding the siting of a new town—and in selecting materials. We turn to that topic next.

8.3 Principles of materials selection

Selecting materials involves seeking the best match between design requirements and the properties of the materials that might be used to make the

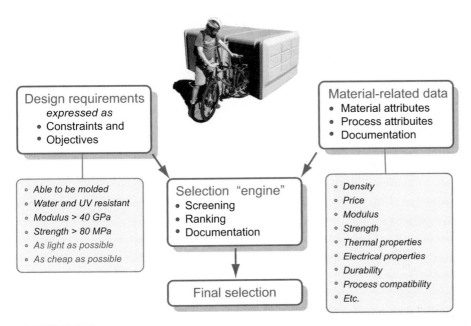

FIGURE 8.3 *Selecting a material for a portable bike shed. The requirements are expressed as constraints and objectives (objectives in blue). Records containing data for materials are screened using the constraints and ranked by the objectives to find the most attractive candidates. These are then explored further by examining documentation.*

product. Figure 8.3 shows the strategy of the last section applied to selecting materials for a portable bike shed. On the left is the list of requirements that the material must meet, expressed as constraints and objectives. The constraints: ability to be molded, weather resistance, adequate stiffness, and strength. The objectives: as light and as cheap as possible. On the right is the database of material attributes, drawn from suppliers' data sheets, handbooks, Web-based sources, or software specifically designed for materials selection. The selection "engine" applies the constraints on the left to the materials on the right and ranks the survivors using an objective, delivering a short list of viable candidates, just as we did with cars. If both objectives are active, tradeoff methods like that of Figure 8.2 resolve the conflict.

There is, however, a complication. The requirements for the car were straightforward—doors, fuel type, power—all explicitly listed by the manufacturer. The design requirements for a component of a product specify what it should *do* but not what *properties* its materials should have. So the first step is one of *translation*: converting the design requirements into constraints and objectives that can be applied to the materials database (see Figure 8.4). The next task is that of *screening*—as with cars, eliminating

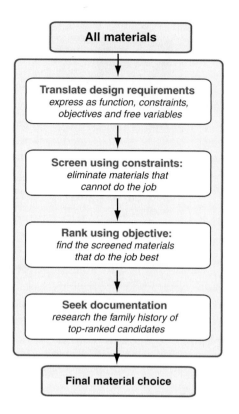

All materials

Translate design requirements
*express as function, constraints,
objectives and free variables*

Screen using constraints:
*eliminate materials that
cannot do the job*

Rank using objective:
*find the screened materials
that do the job best*

Seek documentation
*research the family history of
top-ranked candidates*

Final material choice

FIGURE 8.4   *The strategy. There are four steps: translation, screening, ranking, and supporting information. All can be implemented in software, allowing large populations of materials to be investigated.*

the materials that cannot meet the constraints. This is followed by the *ranking* step, ordering the survivors by their ability to meet a criterion of excellence, such as that of minimizing cost, embodied energy, or carbon footprint. The final task is to explore the most promising candidates in depth, examining how they are used at present, case histories of failures, and how best to design with them, the step we called *documentation*. Now a closer look at each step.

Translation. Any engineering component has one or more *functions*: to support a load, to contain a pressure, to transmit heat, and so forth. This must be achieved subject to *constraints*: that certain dimensions are fixed, that the component must carry the design loads without failure, must insulate against or conduct heat or electricity, must function safely in a certain range of temperature and in a given environment, and many more.

Table 8.1	Function, constraints, objectives, and free variables
Function	What does the component do?
Constraints	What nonnegotiable conditions must be met?
Objective	What is to be maximized or minimized?
Free variables	What parameters of the problem is the designer free to change?

In designing the component, the designer has one or more *objectives*: to make it as cheap as possible, perhaps, or as light, or as environmentally benign, or some combination of these. Certain parameters can be adjusted to optimize the objective; the designer is free to vary dimensions that are not constrained by design requirements and, most important, free to choose the material for the component. We call these *free variables*. Constraints, objectives, and free variables (see Table 8.1) define the boundary conditions for selecting a material and, in the case of load-bearing components, the choice of shape for its cross-section.

It is important to be clear about the distinction between constraints and objectives. A *constraint* is an essential condition that must be met, usually expressed as an upper or lower limit on a material property. An *objective* is a quantity for which an extreme value (a maximum or minimum) is sought, frequently the minimization of cost, mass, volume, or—of particular relevance here—environmental impact (see Table 8.2).

The outcome of the translation step is a list of the design-limiting properties and the constraints they must meet. The first step in relating design requirements to material properties is therefore a clear statement of function, constraints, objectives, and free variables.

Screening. Constraints are gates: meet the constraint and you pass through the gate, fail to meet it and you are shut out. Screening (Figure 8.4) does just that: it eliminates candidates that cannot do the job at all because one or more of their attributes lies outside the limits set by the constraints. As examples, the requirement that "the component must function in boiling water" or that "the component must be nontoxic" imposes obvious limits on the attributes of *maximum service temperature* and *toxicity* that successful candidates must meet. The left column of Table 8.2 lists common constraints.

Ranking: material indices. To rank the materials that survive the screening step we need criteria of excellence—what we have called objectives. The right column of Table 8.2 lists common objectives. Performance is sometimes limited by a single property, sometimes by a combination of them.

Table 8.2	Examples of common constraints and objectives*	
Common constraints		**Common objectives**
Must be:		**Minimize:**
Electrically conducting		*Cost*
Optically transparent		Mass
Corrosion resistant		Volume
Nontoxic		Thermal losses
Nonrestricted substance		Electrical losses
Able to be recycled		*Resource depletion*
		Energy consumption
Must meet a target value of:		*Carbon emissions*
Stiffness		*Waste*
Strength		*Environmental impact*
Fracture toughness		
Thermal conductivity		
Service temperature		

Environment-related constraints and objectives are italicized.

Thus the best materials to minimize thermal losses (an objective) are the ones with the smallest values of the thermal conductivity, λ; those to minimize DC electrical losses (another objective) are those with the lowest electrical resistivity ρ_e—provided, of course, that they also meet all other constraints imposed by the design. Here the objective is met by minimizing a single property. Often, though, it is not one but a group of properties that are relevant. Thus the best materials for a light stiff tie-rod are those with the smallest value of the group, ρ/E, where ρ is the density and E is Young's modulus. Those for a strong beam of lowest embodied energy are those with the lowest value of $H_m\rho/\sigma_y^{2/3}$, where H_m is the embodied energy of the material and σ_y is its yield strength. The property or property-group that maximizes performance for a given design is called its *material index*.

Table 8.3 lists indices for stiffness and strength-limited design for three generic components—a tie-rod, a beam, and a panel—for each of four objectives. The first three relate to design for the environment. Selecting materials with the objective of minimizing volume uses as little materials as possible, conserving resources. Selection with the objective of minimizing mass is central to the ecodesign of transport systems (or indeed of anything that moves) because fuel consumption for transport scales with weight. Selection with the objective of minimizing embodied energy is important when large quantities of material are used, as they are in construction of

Table 8.3	Indices for stiffness and strength-limited design				
Configuration and objective	Configuration	Minimum volume: minimize	Minimum mass: minimize	Minimum embodied energy: minimize	Minimum material cost: minimize
Stiffness-Limited Design	Tie	$1/E$	ρ/E	$H_m\rho/E$	$C_m\rho/E$
	Beam	$1/E^{1/2}$	$\rho/E^{1/2}$	$H_m\rho/E^{1/2}$	$C_m\rho/E^{1/2}$
	Panel	$1/E^{1/3}$	$\rho/E^{1/3}$	$H_m\rho/E^{1/3}$	$C_m\rho/E^{1/3}$
Strength-Limited Design	Tie	$1/\sigma_y$	ρ/σ_y	$H_m\rho/\sigma_y$	$C_m\rho/\sigma_y$
	Beam	$1/\sigma_y^{2/3}$	$\rho/\sigma_y^{2/3}$	$H_m\rho/\sigma_y^{2/3}$	$C_m\rho/\sigma_y^{2/3}$
	Panel	$1/\sigma_y^{1/2}$	$\rho/\sigma_y^{1/2}$	$H_m\rho/\sigma_y^{1/2}$	$C_m\rho/\sigma_y^{1/2}$

Table 8.4	Indices for thermal design		
Objective	Minimum steady-state heat loss: minimize	Minimum thermal inertia: minimize	Minimum heat loss in a thermal cycle: minimize
	λ	$C_p\rho$	$(\lambda C_p\rho)^{1/2}$

Density, ρ (kg/m^3)   Price, C_m ($/kg)

Elastic (Young's) modulus, E (GPa)   Yield strength, σ_y (MPa)

Thermal conductivity λ (W/m.K)   Specific heat, C_p (J/kg.K)

Thermal diffusivity, $a = \lambda/C_p\rho$ (m^2/s)   Embodied energy/kg of material, H_m (MJ/kg)

buildings, bridges, roads, and other infrastructure. The fourth column, selection with the objective of minimizing cost, is always with us.

Table 8.4 lists indices for thermal design. The first is a single property, the thermal conductivity λ; materials with the lowest values of λ minimize heat loss at steady state, that is, when the temperature gradient is constant. The other two guide material choice when the temperature fluctuates. The symbols are defined below Table 8.4.

There are many such indices, each associated with maximizing some aspect of performance. They provide criteria of excellence that allow ranking of materials by their ability to perform well in the given application. Their derivation is described more fully in the appendix to this chapter and in Chapter 9. All can be plotted on material property charts to identify the best candidates. The charts for the indices of Tables 8.3 and 8.4 appear later in this chapter (Section 8.6).

To summarize, then: *screening* uses constraints to isolate candidates that are capable of doing the job; *ranking* uses an objective to identify the candidates that can do the job best.

Documentation. The outcome of the steps so far is a ranked shortlist of candidates that meet the constraints and are ranked most highly by the objective. You could just choose the top-ranked candidate, but what hidden weaknesses might it have? What is its reputation? Has it a good track record? To proceed further we seek a detailed profile of each: its documentation (Figure 8.4, bottom).

What form does documentation take? Typically, it is descriptive, graphical, or pictorial: case studies of previous uses of the material, details of its corrosion behavior in particular environments, of its availability and pricing, warnings of its environmental impact or toxicity, or descriptions of how it is recycled. Such information is found in handbooks, suppliers' data sheets, Websites of environmental agencies, and other high-quality Websites. Documentation helps narrow the shortlist to a final choice, allowing a definitive match to be made between design requirements and material choice.

Why are all these steps necessary? Without screening and ranking, the candidate pool is enormous and the volume of documentation is overwhelming. Dipping into it, hoping to stumble on a good material, gets you nowhere. But once a small number of potential candidates have been identified by the screening-ranking steps, you can seek detailed documentation for these few alone, and the task becomes viable.

8.4 Selection criteria and property charts

Material property charts were introduced in Chapter 6. They are of two types: *bar charts* and *bubble charts*. A bar chart is simply a plot of one or a group of properties; Chapter 6 has several of them. Bubble charts plot two properties or groups of properties. Constraints and objectives can be plotted on them.

Screening: constraints on charts. As we have seen, design requirements impose nonnegotiable demands ("constraints") on the material of which a product is made. These limits can be plotted as horizontal or vertical lines on material property charts. Figures 8.5 and 8.6 show two examples. The first is a bar chart of embodied energy. A selection line has been placed to impose the limit *embodied energy* $< 10 MJ/kg$; all the materials below the line meet the constraint. The second shows a schematic, the *modulus-density* chart. We suppose that the design imposes limits on these of *modulus* $> 10 GPa$ and *density* $< 2000 kg/m^3$, shown on the figure. All materials in the window defined by the limits, labeled "Search region," meet both constraints.

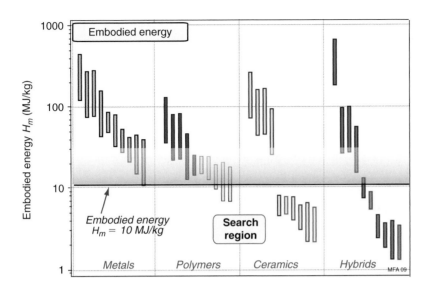

FIGURE 8.5   *Screening using a bar chart. Here we seek materials with embodied energies less than 10 MJ/kg. The materials in the "Search region" below the selection line meet the constraint.*

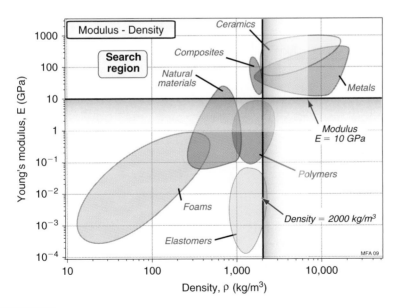

FIGURE 8.6   *Screening using a bubble chart. Two constraints are plotted: modulus >10 GPa and density < 2000 kg/m³. The materials in the "Search region" at the upper left meet both constraints.*

Ranking: indices on charts. Material indices measure performance; they allow ranking of the materials that meet the constraints of the design. We use the design of light, stiff components as examples; the other material indices are used in a similar way.

Figure 8.7 shows a schematic of the $E-\rho$ chart shown earlier. The logarithmic scales allow all three of the indices ρ/E, $\rho/E^{1/3}$ and $\rho/E^{1/2}$, listed in Table 8.3 of the last section, to be plotted onto it. Consider the condition

$$M = \frac{\rho}{E} = constant, C \qquad (8.1)$$

that is, a particular value of the specific stiffness. Taking logs

$$log(E) = log(\rho) - log(C) \qquad (8.2)$$

For a fixed value of C this is the equation of a straight line of slope 1 on a plot of $log(E)$ against $log(\rho)$, as shown in the figure. Similarly, the condition

$$M = \frac{\rho}{E^{1/3}} = constant, C \qquad (8.3)$$

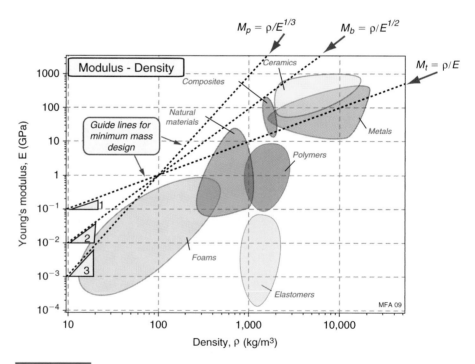

FIGURE 8.7 *A schematic $E-\rho$ chart showing guidelines for three material indices for stiff, lightweight structures.*

becomes, on taking logs,

$$log(E) = 3log(\rho) - 3log(C) \tag{8.4}$$

This is another straight line, this time with a slope of 3, also shown. And by inspection, the third index $\rho/E^{1/2}$ will plot as a line of slope 2. We refer to these lines as *selection guidelines*. They give the slope of the family of parallel lines belonging to that index. Selection guidelines are marked on the charts presented later in this chapter.

It is now easy to read off the subset of materials that maximize performance for each loading geometry. For example, all the materials that lie on a line of constant $M = \rho/E^{1/3}$ perform equally well as a light, stiff panel; those above the line perform better, those below, less well. Figure 8.8 shows a grid of lines corresponding to values of $M = \rho/E^{1/3}$ from $M = 100$ to $M = 10,000$ in units of $(kg.m^{-3})/GPa^{1/3}$. A material with $M = 100$ in these units gives a panel that has one tenth the weight of one with $M = 1000$. The texts listed under Further Reading develop numerous case studies illustrating the use of the method.

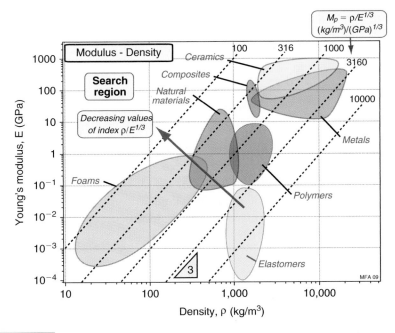

FIGURE 8.8 *A schematic E–ρ chart showing a grid of lines for the index $\rho/E^{1/3}$. The units are $(kg/m^3)/(GPa)^{1/3}/(kg/m^3)$.*

8.5 Resolving conflicting objectives: tradeoff methods

Just as with cars, real-life materials selection almost always requires that a compromise be reached between conflicting objectives. Table 8.2 lists nine of them, and there are more. The choice of materials that best meets one objective will not usually be that which best meets the others; the lightest material, for instance, will generally not be the cheapest or the one with the lowest embodied energy. To make any progress, the designer needs a way of trading weight against cost. This section describes ways of resolving this and other conflicts of objective.

Such conflicts are not new; engineers have sought methods to overcome them for at least a century. The traditional approach is that of using experience and judgment to assign *weight factors* to each constraint and objective, using them to guide choice in the following way.

Weight factors. Weight factors seek to quantify judgment. The method works like this: the key properties or indices are identified and their values M_i are tabulated for promising candidates. Since their absolute values can differ widely and depend on the units in which they are measured, each is first scaled by dividing it by the largest index of its group, $(M_i)_{max}$, so that the largest, after scaling, has the value 1. Each is then multiplied by a weight factor, w_i, with a value between 0 and 1, expressing its relative importance for the performance of the component. This gives a weighted index W_i:

$$W_i = w_i \frac{M_i}{(M_i)_{max}} \tag{8.5}$$

For properties that are not readily expressed as numerical values, such as weldability or wear resistance, rankings such as A to E are expressed instead by a numeric rating, A = 5 (very good) to E = 1 (very bad), then dividing by the highest rating value as before. For properties that are to be minimized, such as corrosion rate, the scaling uses the minimum value $(M_i)_{min}$, expressed in the form

$$W_i = w_i \frac{(M_i)_{min}}{M_i} \tag{8.6}$$

The weight factors w_i are chosen such that they add up to 1, that is: $w_i < 1$ and $\Sigma w_i = 1$. The most important property is given the largest w, the second most important, the second largest, and so on. The W_i's are calculated from Equations 8.5 and 8.6 and summed. The best choice is the material with the largest value of the sum

$$W = \Sigma_i W_i \qquad (8.7)$$

Sounds simple, but there are problems, some obvious (like that of subjectivity in assigning the weights), some more subtle. Experienced engineers can be good at assessing relative weights, but the method nonetheless relies on judgment, and judgments can differ. For this reason, the rest of this section focuses on systematic methods.

Systematic tradeoff strategies. Consider the choice of material to minimize both mass (performance metric P_1) and cost (performance metric P_2) while also meeting a set of constraints such as a required strength or durability in a certain environment. Following the standard terminology of optimizations theory, we define a *solution* as a viable choice of material, meeting all the constraints but not necessarily optimal by either of the objectives. Figure 8.9 is a plot of P_1 against P_2 for alternative solutions, each bubble describing a solution. The solutions that minimize P_1 do not minimize P_2, and vice versa. Some solutions, such as that at A, are far from optimal; all the solutions in the box attached to it have lower values of both P_1 and P_2. Solutions like A are said to be *dominated* by others. Solutions like those at B have the characteristic that no other solutions exist with lower values of

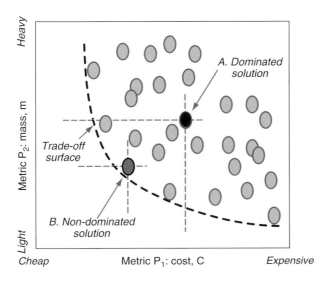

FIGURE 8.9 *Multiple objectives. Mass and cost for a component made from alternative material choices. The tradeoff surface links nondominated solutions.*

both P_1 and P_2. These are said to be *nondominated* solutions. The line or surface on which they lie is called the nondominated or optimal *tradeoff surface*. The values of P_1 and P_2 corresponding to the nondominated set of solutions are called the *Pareto set*.

Just as with cars (Figure 8.2), the solutions on or near the tradeoff surface offer the best compromise; the rest can be rejected. Often this is enough to identify a shortlist, using intuition to rank them. When it is not, the strategy is to define a *penalty function*.

Penalty functions. Consider first the case in which one of the objectives to be minimized is cost, C (units: $), and the other is mass, m (units: kg). We define a locally linear penalty function[1] Z (units: $):

$$Z = C + \alpha m \qquad (8.8)$$

Here α is the change in Z associated with unit increase in m and has the units of $/kg. It is called the *exchange constant*. Rearranging gives:

$$m = -\frac{1}{\alpha}C + \frac{1}{\alpha}Z \qquad (8.9)$$

This defines a linear relationship between m and C that plots as a family of parallel penalty lines, each for a given value of Z, as shown in Figure 8.10. The slope of the lines is the negative reciprocal of the exchange constant, $-1/\alpha$. The value of Z decreases toward the bottom left: the best choices lie there. The optimum solution is the one nearest the point at which a penalty line is tangential to the tradeoff surface, since it is the one with the smallest value of Z.

Values for the exchange constants, α. An exchange constant is the value or "utility" of a unit change in a performance metric. In the example we have just seen, it is the utility ($) of saving 1 kg of weight. Its magnitude and sign depend on the application. Thus the utility of weight saving in a family car is small, though significant; in aerospace it is much larger. The utility of heat transfer in house insulation is directly related to the cost of the energy used to heat the house; that in a heat exchanger for electronics can be much higher because high heat transfer allows faster data processing,

---

[1] Also called a *value function* or *utility function*. The method allows a local minimum to be found. When the search space is large, it is necessary to recognize that the values of the exchange constants α_i may themselves depend on the values of the performance metrics P_i.

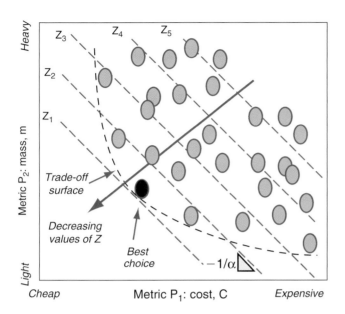

The penalty function Z superimposed on the tradeoff plot. The contours of Z have a slope of $-1/\alpha$. The contour that is tangent to the tradeoff surface identifies the optimum solution.

something worth far more. The utility can be real, meaning that it measures a true saving of cost. But it can also, sometimes, be perceived, meaning that the consumer, influenced by scarcity, advertising, or fashion, will pay more or less than the true value of the performance metric.

In many engineering applications, the exchange constants can be derived approximately from technical models for the life cost of a system. Thus the utility of weight saving in transport systems is derived from the value of the fuel saved or that of the increased payload, evaluated over the life of the system. Table 8.5 gives approximate values for α for various modes of transport. The most striking thing about them is the enormous range: the exchange constant depends in a dramatic way on the application in which the material will be used. It is this that lies behind the difficulty in adopting aluminum alloys for cars, despite their universal use in aircraft; it explains the much greater use of titanium alloys in military than in civil aircraft, and it underlies the restriction of beryllium (a very expensive metal) to use in space vehicles.

Exchange constants can be estimated approximately in various ways. The cost of launching a payload into space lies in the range of $3,000 to $10,000/kg; a reduction of 1 kg in the weight of the launch structure would allow a corresponding increase in payload, giving the ranges of α shown in

Table 8.5	Exchange constants α for the mass-cost tradeoff for transport systems	
Sector: transport systems	**Basis of estimate**	**Exchange constant, α US\$/kg**
Family car	Fuel saving	1–2
Truck	Payload	5–20
Civil aircraft	Payload	100–500
Military aircraft	Payload, performance	500–1000
Space vehicle	Payload	3000–10,000

the table. Similar arguments based on increased payload or decreased fuel consumption give the values shown for civil aircraft, commercial trucks, and automobiles. The values change with time, reflecting changes in fuel costs, legislation to increase fuel economy, and the like.

These values for the exchange constant are based on engineering criteria. More difficult to assess are those based on perceived value. That of the performance/cost tradeoff for cars is an example. To the enthusiast, a car that is able to accelerate rapidly is alluring. He or she is prepared to pay more to go from 0 to 60 mph in 5 seconds than to wait around for 10, as we will see in Chapter 9.

There are other circumstances in which establishing the exchange constant can be more difficult still. An example is that of *environmental impact*—the damage to the environment caused by manufacture, use, or disposal of a given product. Minimizing environmental impact has now become an important objective, almost as important as minimizing cost. Ingenious design can reduce the first without driving the second up too much. But how much is unit decrease in impact worth?

Exchange constants for ecodesign. We explored this topic in Chapter 5. Interventions, as they are called, use taxes, subsidies, and trading schemes to assign a monetary value to resource consumption, energy, emissions, and waste, effectively establishing an exchange constant. As we'll see in the next chapter, none is yet large enough to make big changes in the way we reduce any of these.

8.6 Five useful charts

Five material property charts guide materials selection to minimize mass, total embodied energy, and thermal losses using the indices of Tables 8.3

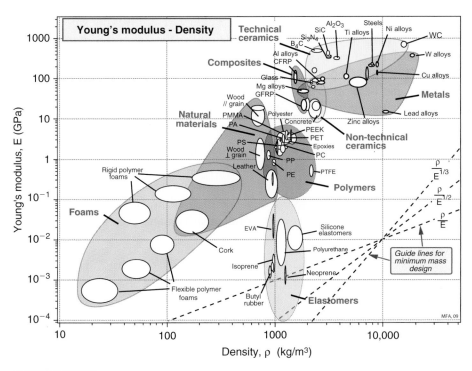

FIGURE 8.11 *The Modulus–Density chart: the one for stiffness at minimum weight.*

and 8.4. They are five of a much larger collection that can be found in the texts listed under Further Reading at the end of this chapter.

■ *The Modulus-Density chart (see Figure 8.11).* The modulus E of engineering materials spans seven decades,[2] from 0.0001 GPa to nearly 1000 GPa; the density ρ spans a factor of 2000, from less than 0.01 to 20 Mg/m^3. Members of each family cluster together and can be enclosed in envelopes, each of which occupies a characteristic part of the chart. The members of the ceramics and metals families have high moduli and densities; none have a modulus less than 10 GPa or a density less than 1.7 Mg/m^3. Polymers, by contrast, all have moduli below 10 GPa and densities that are lower than those of any metal or ceramic; most are close to 1 Mg/m^3. Elastomers have roughly the same density as other polymers, but their moduli are lower by a

---

[2] Very low-density foams and gels (which can be thought of as molecular-scale, fluid-filled foams) can have lower moduli than this. For example, gelatin (as in Jell-O) has a modulus of about 10^{-5} GPa.

further factor of 100 or more. Materials with a lower density than polymers are porous; these include manmade foams and natural cellular structures such as wood and cork.

The three indices for lightweight, stiffness-limited design in Table 8.3 can be plotted onto this chart. The "Guidelines" show the slope associated with each.

■ *The Strength-Density chart (see Figure 8.12).* The range of the yield strength σ_y or elastic limit σ_{el} of engineering materials, like that of the modulus, spans about six decades: from less than 0.01 MPa for foams, used in packaging and energy-absorbing systems, to 10^4 MPa for diamond, exploited in diamond tooling for machining and polishing. Members of each family again cluster together and can be enclosed in envelopes.

Comparison with the modulus-density chart (Figure 8.11) reveals some marked differences. The modulus of a solid is a well-defined quantity with a narrow range of values. The yield strength is not. The strength range for a given class of metals, such as stainless steels, can span a factor

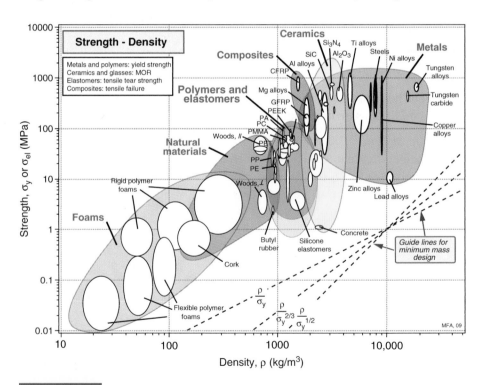

FIGURE 8.12 *The Strength–Density chart: the one for strength at minimum weight.*

of 10 or more, depending on its state of work hardening and heat treatment. Polymers cluster together with strengths between 10 and 100 MPa. The composites CFRP and GFRP have strengths that lie between those of polymers and ceramics, as one might expect, since they are mixtures of the two.

The three indices for lightweight, strength-limited design in Table 8.3 can be plotted onto this chart. The guidelines show the slope associated with each.

- *The Modulus–Embodied Energy and Strength–Embodied Energy charts (see Figures 8.13 and 8.14).* The two charts just described guide design to minimizing mass. If the objective becomes minimizing the energy embodied in the material of the product, we need equivalent charts for these.

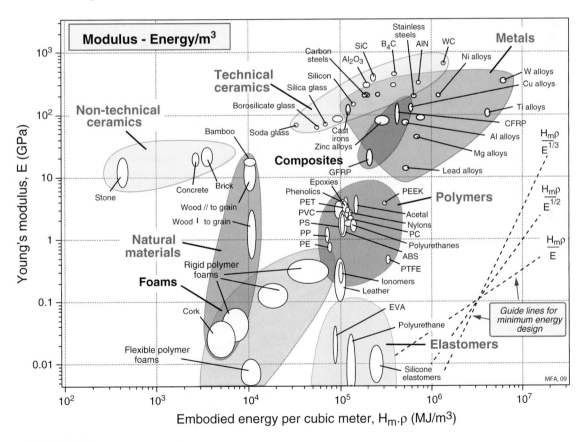

FIGURE 8.13   *The Modulus–Embodied energy chart: the one for stiffness at minimum embodied energy.*

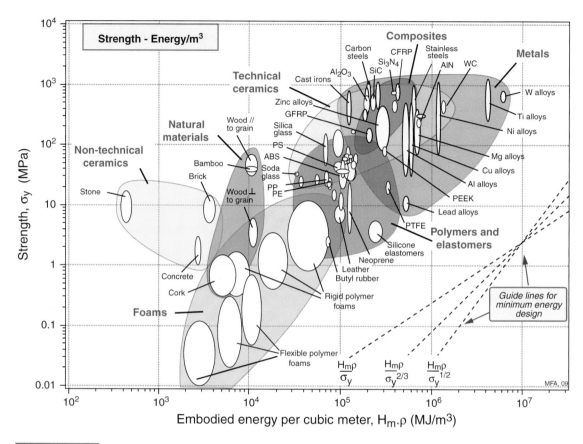

Figure 8.13 shows modulus E plotted against $H_m\rho$; the guidelines give the slopes for three of the most common performance indices for stiffness-limited design at minimum embodied energy. Figure 8.14 shows strength σ_y plotted against $H_m\rho$; again, the guidelines give the slopes for strength-limited design at minimum embodied energy. They are used in exactly the same way as the $E-\rho$ and $\sigma_y-\rho$ charts for minimum mass design.

■ *The Thermal Conductivity–Thermal Diffusivity chart (see Figure 8.15).* Thermal conductivity, λ, is the material property that governs the flow of heat, q (W/m²), in a steady temperature gradient dT/dx:

$$q = -\lambda \frac{dT}{dx} \qquad (8.10)$$

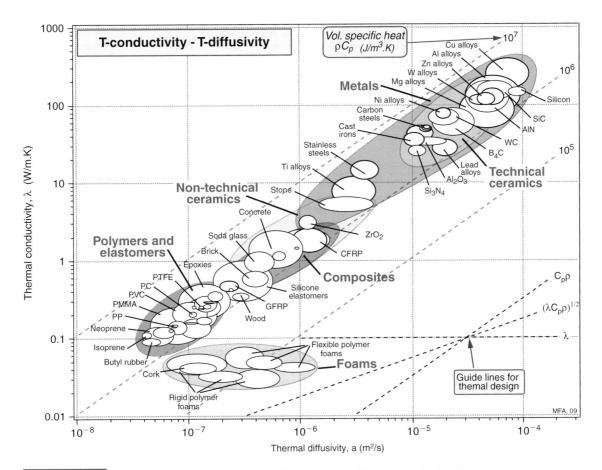

FIGURE 8.15 *The Thermal Conductivity–Thermal Diffusivity chart, with contours of volumetric specific heat: the one for minimum thermal loss.*

The thermal diffusivity, a (m²/s), is the property that determines how quickly heat diffuses into a material. It is related to the conductivity:

$$a = \frac{\lambda}{C_p \rho} \qquad (8.11)$$

where ρC_p is the specific heat per unit mass (J/kg.K). The contours show the volumetric specific heat ρC_p, equal to the ratio of the two, λ/a. The data spans almost five decades in λ and a. Solid materials are strung out along the line

$$\rho C_p \approx 3 \times 10^6 \ \text{J/m}^3.\text{K} \qquad (8.12)$$

meaning that the heat capacity per unit volume, ρC_p is almost constant for all solids, something to remember for later. As a general rule, then,

$$\lambda = 3 \times 10^6 \, a$$

(λ in W/m.K and a in m^2/s). Some materials deviate from this rule: they have lower-than-average volumetric heat capacity. The largest deviations are shown by porous solids: foams, low-density firebrick, woods, and the like. Because of their low density, they contain fewer atoms per unit volume and, averaged over the volume of the structure, ρC_p is low. The result is that, although foams have low *conductivities* (and are widely used for insulation because of this), their *thermal diffusivities* are not necessarily low. This means that they don't transmit much heat, but they do change temperature quickly.

8.7 Computer-aided selection

The charts we've just discussed give an overview, but the number of materials that can be shown on any one of them is obviously limited. Selection using them is practical when there are very few constraints, but when there are many, as there usually are, checking that a given material meets them all is cumbersome. Both problems are overcome by a computer implementation of the method.

The *CES* material selection software[2] is an example of such an implementation. Its database contains records for materials, organized in the hierarchical manner. Each record contains property data for a material, each property stored as a range spanning its typical (or, often, permitted) values. It also contains limited documentation in the form of text, images, and references to sources of information about the material. The data are interrogated by a search engine that offers the search interfaces shown schematically in Figure 8.16.

On the left is a simple query interface for screening on single attributes. The desired upper or lower limits for constrained properties are entered; the search engine rejects all materials with attributes that lie outside the limits. In the center is shown a second way of interrogating the data: a bar chart, constructed by the software, for any numeric property in the database. It, and the bubble chart shown on the right, are ways of both applying constraints and ranking. For screening, a selection line or box is superimposed on the charts with edges that lie at the constrained values

---

[2] The CES material selection software is a product of Granta Design, Cambridge, U.K. (www.grantadesign.com).

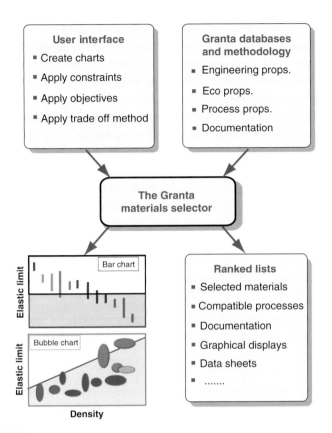

FIGURE 8.16   *The operation and outputs of typical selector software.*

of the property (bar chart) or properties (bubble chart). This eliminates the materials in the shaded areas and retains the materials that meet the constraints. If, instead, ranking is sought (having already applied all necessary constraints), an index line like that shown in Figure 8.8 is positioned so that a small number—say, 10—materials are left in the selected area; these are the top-ranked candidates. The software delivers a ranked list of the top-ranked materials that meet all the constraints.

8.8 Summary and conclusion

There is a broad strategy that works for selecting anything—products, services, or materials. Decide on the attributes the entity must (or must not) have, defining a set of constraints. Apply them, leaving a list of the entities that meet them. Decide on an objective—a measure of excellence.

It could be price (the cheaper, the better), weight (the lighter, the better), or eco-impact (the lower, the better), or some other measure of performance. Use this measure to rank the list of survivors. Then get to work researching the top three or four on the list, gathering as much information as possible to make a well-informed final choice.

There are, almost always, many constraints, but this does not create a difficulty: simply apply them sequentially, retaining only those entities that meet them all. Often, too, there are two or more objectives, and that does create a difficulty; the entity that best satisfies one is unlikely to also be the best choice by the other. Then tradeoff methods become useful, either graphical ones (plotting the alternatives, identifying the tradeoff line, and then using judgment to select an entity on or near the line) or analytical ones, by formulating a penalty function and seeking the entitites that carry the lowest penalty.

Now we have a set of tools. In Chapters 9 these tools will be used to analyze and select materials to design for the environment.

8.9 Further reading

Ashby, M.F. (2005), "Materials selection in mechanical design", 3rd ed., Butterworth Heinemann. Chapter 4, ISBN 0-7506-6168-2. *(A text that develops the ideas presented here in more depth, including the derivation of material indices, a discussion of shape factors, and a catalog of simple solutions to standard problems.)*

Ashby, M.F., Shercliff, H.R. and Cebon, D. (2007), "Materials: engineering, science, processing and design", Butterworth Heinemann. ISBN-13: 978-0-7506-8391-3. *(An elementary text introducing materials through material property charts and developing the selection methods through case studies.)*

Bader, M.G. Proc of ICCM-11, Gold Coast, Australia, Vol. 1: Composites applications and design, ICCM, 1977. *(An example of tradeoff methods applied to the choice of composite systems.)*

Bourell, D.L., Decision matrices in materials selection, ASM Handbook Vol. 20, Materials selection and design, G. E. Dieter (Ed.), ASM International, 1997, 291–296, ISBN 0-87170-386-6. *(An introduction to the use of weight factors and decision matrices.)*

Dieter, G.E. (2000), "Engineering design, a materials and processing approach", 3rd ed., McGraw-Hill. 150–153 and 255–257, ISBN 0-07-366136-8. *(A well-balanced and respected text, now in its third edition, focusing on the role of materials and processing in technical design.)*

Field, F.R., and Neufville, R. de, "Material selection: maximizing overall utility, Metals and Materials", June 1998, 378–382. *(A summary of utility analysis applied to material selection in the automobile industry.)*

Goicoechea, A., Hansen, D.R. and Druckstein, L. (1982), "Multi-objective decision analysis with engineering and business applications", Wiley. *(A good starting point for the theory of multiobjective decision making.)*

Keeney, R.L. and Raiffa, H. (1993), "Decisions with multiple objectives: preferences and value tradeoffs", 2nd ed., Cambridge University Press. ISBN 0-521-43883-7. (*A notably readable introduction to methods of decision making with multiple, competing objectives.*)

8.10 Appendix: deriving material indices

This appendix describes how material indices are derived. You can find out more about them and their use in the first two texts listed under Further Reading.

The performance of a component is characterized by a performance equation called the *objective function*. The performance equation contains a group of material properties. This group is the material indices of the problem. Sometimes the "group" is a single property; thus if the performance of a beam is measured by its stiffness, the performance equation contains only one property, the elastic modulus E. More commonly, the performance equation contains a group of two or more properties. Familiar examples are the *specific stiffness*, E/ρ, (where E is Young's modulus and ρ is the density), and the *specific strength*, σ_y/ρ (where σ_y is the yield strength or elastic limit), but there are many others. For reasons that will become apparent, we express the indices in a form for which a *minimum*, not a *maximum*, is sought.

Recall that the life energy and emissions for transport systems are dominated by the fuel consumed during use. The lighter the system is made, the less fuel it consumes and the less carbon it emits. So a good starting point is *minimum weight design*, subject, of course, to the other necessary constraints of which the most important here have to do with stiffness and strength. We consider the generic components shown in Figure 8.17: ties, panels, and beams, loaded as shown. The derivation when the objective is that of minimizing embodied energy or material cost follows in a similar way.

Minimizing mass: a light, stiff tie rod. A material is sought for a cylindrical tie rod that must be as light as possible (Figure 8.17a). Its length L_o is specified and it must carry a tensile force F without extending elastically by more than δ. Its stiffness must be at least $S^\star = F/\delta$. We are free to choose the cross-section area A and, of course, the material. The design requirements, translated, are listed in Table 8.6.

We first seek an equation that describes the quantity to be minimized, here the mass m of the tie. This equation, the *objective function*, is

$$m = A L_o \rho \qquad (8.13)$$

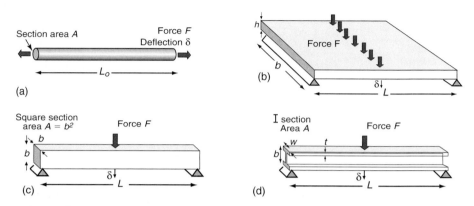

FIGURE 8.17   Generic components. (a) A tie, a tensile component; (b) a panel, loaded in bending; (c) and (d) beams, loaded in bending.

Table 8.6	Design requirements for the light, stiff tie rod
Function	**Tie rod**
Constraints	Stiffness S^* specified (a functional constraint)
	Length L_0 specified (a geometric constraint)
Objective	Minimize mass
Free variables	Choice of material
	Choice of cross-section area A

where ρ is the density of the material of which it is made. We can reduce the mass by reducing the cross-section, but there is a constraint: the section-area A must be sufficient to provide a stiffness of S^*, which, for a tie, is:

$$S = \frac{A\,E}{L_0} \geq S^\star \qquad (8.14)$$

where E is Young's modulus. If the material has a low modulus, a large A is needed to give the necessary stiffness; if E is high, a smaller A is needed. But which gives the lower mass? To find out, we eliminate the free variable A between these two equations, giving

$$m = S^\star L_0^2 \left(\frac{\rho}{E}\right) \qquad (8.15)$$

Table 8.7	Design requirements for the light, stiff panel
Function	**Panel**
Constraints	Stiffness S^* specified (a functional constraint)
	Length L and width b specified (a geometric constraint)
Objective	Minimize mass
Free variables	Choice of material
	Choice of panel thickness h

Both S^* and L_o are specified. The lightest tie that will provide a stiffness $S^{\cdot}$ is that made of the material with the smallest value of the index

$$M_{t_1} = \frac{\rho}{E} \qquad (8.16a)$$

provided that they also meet all other constraints of the design. If the constraint is not stiffness but strength, the index becomes

$$M_{t_2} = \frac{\rho}{\sigma_y} \qquad (8.16b)$$

where σ_y is the yield strength. That means that the best choice of material for the lightest tie that can support a load F without yielding is that with the smallest value of this index.

The mode of loading that most commonly dominates in engineering is not tension but bending—think of floor joists of buildings, wing spars of aircraft, or shafts of golf clubs and racquets. The index for bending differs from that for tension, and this (significantly) changes the optimal choice of material. We start by modeling a panel, specifying stiffness and seeking to minimize its embodied energy.

Minimizing mass: a light, stiff panel. A panel is a flat slab, like a tabletop. Its length L and width b are specified, but its thickness h is free. It is loaded in bending by a central load F (Figure 8.17b). The stiffness constraint requires that it must not deflect more than δ. The objective is to achieve this with minimum mass, m. Table 8.7 summarizes the design requirements.

The objective function for the mass of the panel is the same as that for the tie:

$$m = AL\rho = bhL\rho$$

Its bending stiffness S must be at least $S^\star$:

$$S = \frac{C_1 E I}{L^3} \geq S^\star \tag{8.17}$$

Here C_1 is a constant that depends only on the distribution of the loads and I is the second moment of area, which, for a rectangular section, is

$$I = \frac{b h^3}{12} \tag{8.18}$$

We can reduce the mass by reducing h, but only so far that the stiffness constraint is no longer met. Using the last two equations to eliminate h in the objective function gives

$$m = \left(\frac{12\,S^\star}{C_1 b}\right)^{1/3} \left(b L^2\right) \left(\frac{\rho}{E^{1/3}}\right) \tag{8.19}$$

The quantities $S^\star$, L, b, and C_1 are all specified; the only freedom of choice left is that of the material. The best materials for a light, stiff panel are those with the smallest values of

$$M_{P_1} = \frac{\rho}{E^{1/3}} \tag{8.20a}$$

Repeating the calculation with a constraint of strength rather than stiffness leads to the index

$$M_{P_2} = \frac{\rho}{\sigma_y^{1/2}} \tag{8.20b}$$

These don't look much different from the previous indices, ρ/E and ρ/σ_y but they are; they lead to different choices of material, as we shall see in a moment. For now, note the procedure. The in-plane dimensions of the panel were specified, but we were free to vary the thickness h. The objective is to minimize its mass, m. Use the stiffness constraint to eliminate the free variable, here h. Then read off the combination of material properties that appears in the objective function—the equation for the mass. It sounds easy, and it is—as long as you are clear from the start what the constraints are, what you are trying to maximize or minimize, and which parameters are specified and which are free.

Now for another bending problem, in which the freedom to choose shape is rather greater than for the panel.

Table 8.8	Design requirements for the light, stiff beam
Function	**Beam**
Constraints	Stiffness S^* specified (a functional constraint) Length L (geometric constraints) Section shape square
Objective	Minimize mass
Free variables	Choice of material Area A of cross-section

Minimizing mass: a light, stiff beam. Beams come in many shapes: solid rectangles, cylindrical tubes, I-beams, and more. Some of these have too many free geometric variables to directly apply the previous method. However, if we constrain the shape to be *self-similar* (such that all dimensions change in proportion as we vary the overall size), the problem becomes tractable again. We therefore consider beams in two stages: first, to identify the optimum materials for a light, stiff beam of a prescribed simple shape (a square section); second, we explore how much lighter it could be made, for the same stiffness, by using a more efficient shape.

Consider a beam of square section $A = b \times b$ that may vary in size, but the square shape is retained. It is loaded in bending over a span of fixed length L with a central load F (Figure 8.17c). The stiffness constraint is again that it must not deflect more than δ under the load F, with the objective that the beam should again be as light as possible. Table 8.8 summarizes the design requirements.

Proceeding as before, the objective function for the embodied energy is:

$$m = A L \rho = b^2 L \rho$$

The bending stiffness S of the beam must be at least S^*:

$$S = \frac{C_1 E I}{L^3} \geq S^* \qquad (8.21)$$

where C_1 is a constant; we don't need is value. The second moment of area, I, for a square section beam is

$$I = \frac{b^4}{12} = \frac{A^2}{12} \qquad (8.22)$$

For a given length L, the stiffness $S^\star$ is achieved by adjusting the size of the square section. Now eliminating b (or A) in the objective function for the mass gives

$$m = \left(\frac{12\,S^\star L^3}{C_1}\right)^{1/2} (L) \left(\frac{\rho}{E^{1/2}}\right) \tag{8.23}$$

The quantities $S^\star$, L, and C_1 are all specified or constant; the best materials for a light, stiff beam are those with the smallest values of the index M_b, where

$$M_{b_1} = \frac{\rho}{E^{1/2}} \tag{8.24a}$$

Repeating the calculation with a constraint of strength rather than stiffness leads to the index

$$M_{b_2} = \frac{\rho}{\sigma_y^{2/3}} \tag{8.24b}$$

This analysis was for a square beam, but the result in fact holds for any shape, so long as the shape is held constant. This is a consequence of Equation 8.21: for a given shape, the second moment of area I can always be expressed as a constant times A^2, so changing the shape merely changes the constant C_1 in Equation 8.23, not the resulting index.

As noted, real beams have section shapes that improve their efficiency in bending, requiring less material to get the same stiffness. By shaping the cross-section, it is possible to increase I without changing A. This is achieved by locating the material of the beam as far from the neutral axis as possible, as in thin-walled tubes or I-beams (Figure 8.17d). Some materials are more amenable than others to being made into efficient shapes. Comparing materials on the basis of the index in M_b therefore requires some caution: materials with lower values of the index may "catch up" by being made into more efficient shapes. So we need to get an idea of the effect of shape on bending performance.

Figure 8.18 shows a solid square beam of cross-section area A. If we turn the same area into a tube, as shown on the right of the figure, the mass of the beam is unchanged. The second moment of area, I, however, is now much greater—and so is the stiffness (Equation 8.21). We define the ratio of I for the shaped section to that for a solid square section with the same area (and thus mass) as the *shape factor* Φ. The more slender the shape,

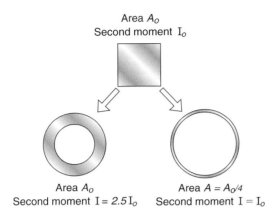

Area A_O
Second moment I_O

Area A_O
Second moment $I = 2.5 I_O$

Area $A = A_O/4$
Second moment $I = I_O$

FIGURE 8.18 *The effect of section shape on bending stiffness EI: a square section beam compared, left, with a tube of the same area (but 2.5 times stiffer) and, right, a tube with the same stiffness (but 4 times lighter).*

Table 8.9	The cffect of shaping on stiffness and mass of beams in different structural materials	
Material	**Typical maximum shape factor (stiffness relative to that of a solid square beam)**	**Typical mass ratio by shaping (relative to that of a solid square beam)**
Steels	64	1/8
Al alloys	49	1/7
Composites (GFRP, CFRP)	36	1/6
Wood	9	1/3

the larger is Φ, but there is a limit—make it too thin and the flanges will buckle—so there is a maximum shape factor for each material that depends on its properties. Table 8.9 lists some typical values.

Shaping is used to make structures lighter; it is a way to get the same stiffness with less material. The mass ratio is given by the reciprocal of the square root of the maximum shape factor, $\Phi^{-1/2}$ (because C_1, which proportional to the shape factor, appears as $(C_1)^{-1/2}$ in Equation 8.23). Table 8.9 lists the factors by which a beam can be made lighter, for the same stiffness, by shaping. Metals and composites can all be improved significantly (though the metals do a little better), but wood has more limited potential because it is more difficult to shape it into efficient, thin-walled shapes. So, when we compare materials for light, stiff beams using the index in Equation 8.24, we find that the performance of wood is not as good as it looks, because other materials can be made into more efficient shape. Composites (particularly CFRP) have attractive (i.e., low) values of all the indices M_t, M_p, and M_b, but this advantage relative to metals is reduced a little by the effect of shape.

Minimizing embodied energy. When the objective is to minimize embodied energy rather than mass, the indices change. If the embodied energy of the material is H_m MJ/kg, the energy embodied in a component of mass m is just mH_m. The objective function for the energy H embodied in the tie, panel, or beam then becomes

$$H = mH_m = ALH_m\rho \qquad (8.25)$$

Proceeding along the same steps as for minimum mass then leads to indices that have the form of Equations 8.16, 8.20, and 8.24, with ρ replaced by $H_m\rho$, as in Table 8.3.

Minimizing material cost. When, instead, the objective is to minimize cost rather than mass, the indices change again. If the material price is C_m $/kg, the cost of the material to make a component of mass m is just mC_m. The objective function for the material cost C of the tie, panel, or beam then becomes

$$C = mC_m = ALC_m\rho \qquad (8.26)$$

Proceeding as before then leads to indices that have the form of Equations 8.16, 8.20, and 8.24 with ρ replaced by $C_m\,\rho$, as in Table 8.3. (It must be remembered that the material cost is only part of the cost of a shaped component; there is also the manufacturing cost—the cost to shape, join, and finish it.)

8.11 Exercises

E.8.1. What is meant by an *objective* and what by a *constraint* in the requirements for a design? How do they differ?

E.8.2. Describe and illustrate the Translation step of the material selection strategy.

E.8.3. Bikes come in many forms, each aimed at a particular sector of the market:

- Sprint bikes
- Touring bikes

■ Mountain bikes
■ Shopping bikes
■ Children's bikes
■ Folding bikes

Use your judgment to identify the primary objective and the constraints that must be met for each of these.

E.8.4. You are asked to design a fuel-saving cooking pan with the goal of wasting as little heat as possible while cooking. What objective would you choose, and what constraints would you recommend must be met?

E.8.5. Formulate the constraints and objective you would associate with the choice of material to make the forks of a racing bicycle.

E.8.6. What is meant by a *material index*?

E.8.7. The objective in selecting a material for a panel of given in-plane dimensions for the lid casing of an ultrathin portable computer is that of minimizing the panel thickness h while meeting a constraint on bending stiffness, S^*, to prevent damage to the screen. What is the appropriate material index?

E.8.8. Plot the index for a light, stiff panel on a copy of the *modulus/density* chart of Figure 8.11, positioning the line such that six materials are left above it, excluding ceramics because of their brittleness. Which six do you find? To what material classes do they belong?

E.8.9. Panels are needed to board up the windows of an unused building. The panels should have the lowest possible embodied energy but be strong enough to deter an intruder who, in attempting to break in, will load the panels in bending. Which index would you choose to guide choice?

Plot the index on the *strength/embodied energy* chart of Figure 8.14, positioning the line to find the best choice, excluding ceramics because of their brittleness. Which six do you find? To what material classes do they belong?

E.8.10. A material is required for a disposable fork for a fast-food chain. List the objective and the constraints that you would see as important in this application.

E.8.11. A designer seeks a material for a disposable drinking cup, with a goal of minimizing the embodied energy. When held between fingers and thumb, the cylindrical cup deflects like a panel in bending. Plot the appropriate index onto the *modulus/embodied energy* and the *strength/embodied energy* chart, using common sense to apply other necessary constraints, and make a selection.

E.8.12. Show that the index for selecting materials for a strong panel with the dimensions shown in Figure 8.17c, loaded in bending, with the minimum embodied energy content, is

$$M = \frac{H_m \rho}{\sigma_y^{1/2}}$$

To do so, rework the panel derivation in Section 8.10, replacing the stiffness constraint with a constraint on failure load F requiring that it exceed a chosen value F^*, where

$$F = C_2 \frac{I \sigma_y}{hL} > F^*$$

where C_2 is a constant and the other symbols have the meaning used in the text.

E.8.13. Use the chart $E - H_m \rho$ of Figure 8.13 to find the metal with a modulus E greater than 100 GPa and the lowest embodied energy per unit volume.

E.8.14. A maker of polypropylene (PP) garden furniture is concerned that the competition is stealing part of his market by claiming that the "traditional" material for garden furniture, cast iron, is much less energy and CO_2 intensive than the PP. A typical PP chair weighs 1.6 kg; one made of cast iron weighs 11 kg. Use the data sheets for these two materials in Chapter 12 of the book to find out who is right. Remember the warning about precision at the start of Chapter 12.

If the PP chair lasts five years and the cast iron chair lasts 25 years, does the conclusion change?

Exploring design using CES Edu level 2 ECO

E.8.15. Use a Limit stage to find materials with modulus $E > 180$ GPa and embodied energy Hm < 30 MJ.kg.

E.8.16. Use a Limit stage to find materials with yield strength $\sigma_y > 100$ MPa and a carbon footprint $CO_2 > 1$ kg/kg.

E.8.17. Make a bar chart of embodied energy H_m. Add a tree stage to limit the selection to polymers alone. Which polymers have the lowest embodied energy?

E.8.18. Make a chart showing modules E and density ρ. Apply a selection line of slope 1, corresponding to the index ρ/E, positioning the line such that six materials are left above it. To what families do they belong?

E.8.19. A material is required for a tensile tie-rod to link the front and back walls of a barn to stabilize both. It must meet a constraint on strength and have as low an embodied energy as possible. To be safe, the material of the tie must have a fracture toughness $K_{1c} > 18\,\text{MPa.m}^{1/2}$. The relevant index is

$$M = H_m \rho \neq \sigma_y$$

Construct a chart of σ_y plotted against $H_m\,\rho$. Add the constraint of adequate fracture toughness, meaning $K_{1c} > 18\,\text{MPa.m}^{1/2}$, using a Limit stage. Then plot an appropriate selection line on the chart and report the three materials that are the best choices for the tie.

E.8.20. A company wants to enhance its image by replacing oil-based plastics in its products with polymers based on natural materials. Use the Search facility in CES to find *biopolymers*. List the materials you find. Are their embodied energies and CO_2 footprints less than those of conventional plastics? Make bar charts of embodied energy and CO_2 footprint to find out.

Eco-informed materials selection

9.1 Introduction and synopsis

Audits like those of Chapter 7 point the finger, directing attention to the life phase that is of most ecoconcern. If you point fingers, you invite the response: what do you propose to do about it? That means moving from *auditing* and assessment to *selection*—from the top part of the strategy of Figure 3.11 to the bottom.

Chapter 8 introduced the methods. Here we illustrate their use with case studies. The first two are simple, showing how selection methods work. Those that follow get progressively more complex, leading up to the last two, which deal with *heating and cooling* and with *transport,* two of the biggest sinks of energy and sources of emissions in industrialized society.

Super lightweight vehicles: shell ecomarathon contester, Cal Poly Supermilage team vehicle, "Microjoule" by the students of the Lycee La Joliverie, France, and "Pivo2" electric car by Nissan. All have shell bodies made from materials chosen for stiffness and strength at minimum mass.

Before starting, there's something to bear in mind. There are no simple, single-answer solutions to environmental questions. Material substitution guided by eco-objectives is one way forward, but it is not the only one. It might sometimes be better to abandon one way of doing things (the IC engine vehicle, for example) and replacing it with another (fuel cell or electric power, perhaps). So, though change of material is one option, another is change of concept. And of course there is a third: change of lifestyle (no vehicle at all).

This book is about materials so, in Chapters 1 through 8, we stuck with them as the central theme. In this and the next two chapters we venture a little outside this envelope.

9.2 Which bottle is best? selection per unit of function

Drink containers coexist that are made from many different materials: glass, polyethylene, PET, aluminum, steel—Figure 9.1 shows them. Surely one must be a better environmental choice than the others? The audit of a PET bottle in Chapter 7 delivered a clear message: the phase of life that dominates energy consumption and CO_2 emission is that embodied in the material of which a product is made. Embodied energies for the five materials are plotted in the upper part of Figure 9.2 (a plot of CO_2 shows the same distribution). Glass has values of both that are by far the lowest. It would seem that glass is the best choice.

But hold on. These are energies *per kg of material.* The containers differ greatly in weight and volume. What we need are values *per unit of function.* So let's start again and do the job properly, listing the design requirements. The material must not corrode in mildly acidic (fruit juice) or alkali (milk) fluids. It must be easy to shape, and—given the short life of a container—it must be recyclable. Table 9.1 lists the requirements, including the objective of minimizing embodied energy *per unit volume of fluid contained.*

Glass   PE   PET   Aluminum   Steel

FIGURE 9.1 *Containers for liquids: glass, polyethylene, PET, aluminum, and steel; all can be recycled. Which carries the low penalty of embodied energy?*

The masses of five competing container types, the material of which they are made, and the embodied energy of each are listed in Table 9.2. All five materials can be recycled. For all five, cost-effective processes exist for making containers. All but one—steel—resist corrosion in the mildly acidic or alkaline conditions characteristic of bottled drinks. Steel is easily protected with lacquers.

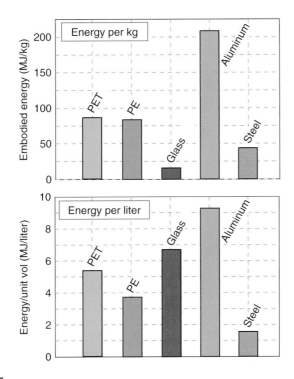

FIGURE 9.2 *Top: the embodied energy of the bottle materials. Bottom: the material energy per liter of fluid contained.*

Table 9.1	Design requirements for drink containers
Function	**Drink container**
Constraints	Must be immune to corrosion in the drink
	Must be easy and fast to shape
	Must be recyclable
Objective	Minimize embodied energy per unit capacity
Free variables	Choice of material

Table 9.2	Data for the containers with embodied energies for virgin material			
Container type	Material	Mass, Grams	Embodied energy MJ/kg	Energy/Liter MJ/Liter
PET 400 ml bottle	PET	25	84	5.3
PE 1 liter milk bottle	High-density PE	38	81	3.8
Glass 750 ml bottle	Soda glass	325	15.5	6.7
Al 440 ml can	5000 series Al alloy	20	208	9.5
Steel 440 ml can	Plain carbon steel	45	32	**3.3**

That leaves us with the objective. The last column of the table lists embodied energies per liter of fluid contained, calculated from the numbers in the other columns of the table. The results are plotted in the lower part of Figure 9.2. The ranking is now very different: steel emerges as the best choice, polythene the next best. Glass (because so much is used to make one bottle) and aluminum (because of its high embodied energy) are the least good.

Postscript: In all discussion of this sort, there are issues of primary and of secondary importance. There is cost; we have ignored this because ecodesign was the prime objective. There is ease of recycling; the value of recycled materials depends on differing degrees on impurity pickup. There is the fact that real cans and bottles are made with some recycled content, reducing embodied energy of all five to varying degrees but not enough to change the ranking. There is the extent to which current legislation subsidizes or penalizes one material or another. And there is appearance: transparency is attractive for some products but irrelevant for others. However, we should not let these cloud the primary finding: that the containers differ in their life energy, dominated by material, and that steel is by far the least energy intensive.

9.3 Crash barriers: matching choice to purpose

Barriers to protect drivers and passengers of road vehicles are of two types: those that are static (the central divider of a freeway, for instance) and those that move (the bumper of the vehicle itself), as shown in Figure 9.3 and Table 9.3. The static type lines tens of thousands of miles of road. Once in place they consume no energy, create no CO_2, and last a long time. The dominant phases of their life in the sense of the life cycle of Figure 3.1 are those of material production and manufacture. The bumper, by contrast, is

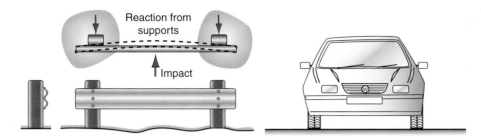

FIGURE 9.3 *Two crash barriers, one static, the other—the bumper—attached to something that moves. Different ecocriteria are needed for each. The barrier is loaded in bending, as in the plan view in the figure.*

Table 9.3	Design requirements for crash barriers
Function	**Crash barrier: transmit impact load to absorbing elements**
Constraint	High strength Adequate fracture toughness Recyclable
Objectives	Minimize embodied energy for given bending strength (static barrier) Minimize mass for a given bending strength (mobile barrier)
Free variables	Choice of material Shape of cross-section

part of the vehicle; it adds to its weight and thus to its fuel consumption. The audit of Chapter 7 established that the dominant phase here is that of use. If ecodesign is the objective, the criteria for selecting materials for the two sorts of barrier will differ: minimizing embodied energy for the first, minimizing mass for the second.

In an impact, the barrier is loaded in bending (Figure 9.3). Its function is to transfer load from the point of impact to the support structure, where reaction from the foundation or from crush elements in the vehicle supports or absorbs it. To do this the material of the barrier must have high strength, σ_y, be adequately tough, and able to be recycled. That for the static barrier must meet these constraints with *minimum embodied energy* as the objective, since this will reduce the overall life energy most effectively. We know from Chapter 8 that this means materials with low values of the index

$$M_1 = \frac{H_m \rho}{\sigma_y^{2/3}} \qquad (9.1)$$

where σ_y is the yield strength, ρ the density, and H_m the embodied energy per kg of material. For the car bumper it is mass, not embodied energy, that is the problem. If we change the objective to that of *minimum mass*, we require materials with low values of the index

$$M_2 = \frac{\rho}{\sigma_y^{2/3}} \qquad (9.2)$$

These indices can be plotted onto the charts of Figures 8.13 and 8.15; we leave that as one of the exercises at the end of this chapter to show here an alternative: simply plotting the index itself as a bar chart. Figures 9.4 and 9.5 show the result for metals, polymers, and polymer-matrix composites. The first guides the selection for static barriers. It shows that embodied energy (for a given load-bearing capacity) is minimized by making the barrier from carbon steel or cast iron or wood; nothing else comes close. The second figure guides selection for the mobile barrier. Here carbon fiber reinforced polymer (CFRP), for instance, excels in its strength per unit weight, but it is not recyclable. Heavier, but recyclable, are alloys of magnesium, titanium, and aluminum. Polymers, which rank poorly on the first figure, now become candidates; even without reinforcement, they can be as good as steel.

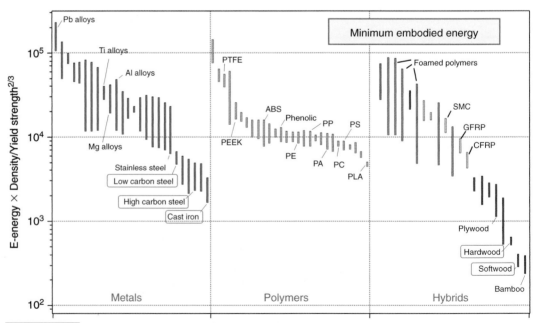

FIGURE 9.4 *Material choice for the static barrier is guided by the bending strength per unit of embodied energy, $H_m \rho / \sigma_y^{2/3}$ (here in units of $(MJ/m^3)/MPa^{2/3}$). Cast irons, carbon steels, low alloy steels, or wood are the best choices. (Here the number of materials has been limited for clarity).*

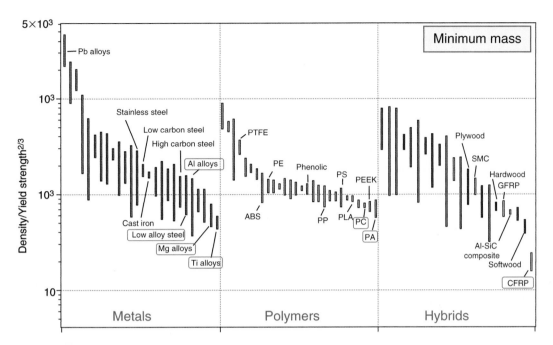

FIGURE 9.5 *Material choice for the mobile barrier is guided by the bending strength per unit of weight, $\rho/\sigma_y^{2/3}$ (here in units of $(kg/m^3)/MPa^{2/3}$). CFRP and light alloys offer the best performance; nylon and polycarbonate (PC) perform better than steel. (Here the number of materials has been limited for clarity).*

Postscript: Metal crash barriers have a profile like that shown on the left of Figure 9.3. The curvature increases the second moment of area of the cross-section, and through this, the bending stiffness and strength. This is an example of combining material choice and section shape (Section 8.10 and Table 8.9) to optimize a design. A full explanation of the coselection of material and shape can be found in the first text listed in Further Reading.

9.4 Deriving and using indices: materials for light, strong shells

The four ecocars pictured on the title page of this chapter all have casings that are thin, doubly curved sheets, or *shells*. The designers wanted a casing that was adequately stiff and strong and as light as possible. The double curvature of the shell helps with this: a shell, when loaded in bending, is stiffer and stronger than a flat or singly curved sheet of the same thickness, because any attempt to bend it creates *membrane stresses*—tensile or compressive stress in the plane of the sheet. Sheets support tension or compression much better than they support bending.

So what is the best material for an adequately stiff and strong shell that is as light as possible? To answer that we need material indices for shells.

Modeling: indices for shells. Figure 9.6 shows a hemispherical shell of radius R and thickness t carrying a distributed load F. The load induces a deflection δ and a maximum membrane stress σ. Define the stiffness S as F/δ. The stiffness constraint then becomes $S \geq S^\star$, where $S^\star$ is the desired stiffness. The strength constraint is simply $\sigma \leq \sigma_y$, where σ_y is the yield strength of the material. Table 9.4 summarizes the requirements.

The mass of the hemisphere (the objective function) is

$$m = 2\pi R^2 t\rho \tag{9.3}$$

where ρ is the density of its material. The deflection and membrane stress σ created by a distributed load like that in the figure are standard results.[1] Using them, the stiffness S is

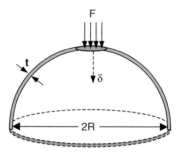

FIGURE 9.6   *A shell, loaded externally with a distributed load F.*

Table 9.4	Design requirements for the light stiff, strong shell	
Function	**Doubly curved shell**	
Constraints	Stiffness S^* specified Failure load F^* specified	(functional constraints)
	Radius R specified Load distribution specified	(geometric constraints)
Objective	Minimize mass	
Free variables	Thickness of shell wall, t Choice of material	

[1]See the compilation by Young listed under Further Reading.

$$S = \frac{F}{\delta} = \frac{Et^2}{AR(1 - \nu)} \geq S^\star \qquad (9.4)$$

and the maximum membrane stress σ is

$$\sigma = B\frac{F}{t^2} \leq \sigma_y \qquad (9.5)$$

where E is Young's modulus, ν is Poisson's ratio, and A and B are constants that depend weakly on how the load is distributed on the surface of the hemisphere. Poisson's ratio is almost the same for all structural materials and can be treated as a constant. Solving each of these for t and substituting the result into the objective function gives

$$m = 2\pi R^2 (AS^\star R(1 - \nu))^{1/2}\left[\frac{\rho}{E^{1/2}}\right] \quad (\textit{stiffness constraint}) \qquad (9.6)$$

and

$$m = 2\pi R^2 (BF)^{1/2}\left[\frac{\rho}{\sigma_y^{1/2}}\right] \quad (\textit{strength constraint}) \qquad (9.7)$$

Everything in these two equations is specified except for the material properties in square brackets, so the two indices are

$$M_1 = \frac{\rho}{E^{1/2}} \qquad (9.8)$$

and

$$M_2 = \frac{\rho}{\sigma_y^{1/2}} \qquad (9.9)$$

The mass of the shell is proportional to the value of the index.

The Selection. Materials for shells can be compared by evaluating the indices or by plotting them onto appropriate charts. Take the index for adequate strength and minimum weight, M_2, as an example. Taking logs of Equation 9.9 and rearranging gives

$$Log\ \sigma_y = 2log\ \rho - 2log\ M_2$$

This is the equation of a family of contours of slope 2 on the $\sigma_y - \rho$ chart of Chapter 8. Figure 9.7 shows this chart with the selection line of slope 2 positioned to leave the three materials with the lowest values of M_2 exposed in the search area. They are CFRP and two grades of rigid polymer foam; certain ceramics come close. A similar selection line for M_1, plotted on the modulus-density chart of Figure 8.11, gives the same result. Ceramics are ruled out by their brittleness. Foams are eliminated for a different reason: they are very light, but to achieve the necessary stiffness and strength, a foam shell has to be thick, increasing the frontal area and drag of the car. That leaves CFRP as the unambiguous best choice.

Postscript. CFRP is what the mileage-marathon cars use. But there is more to it than that. The lowest mass is achieved by a combination of material

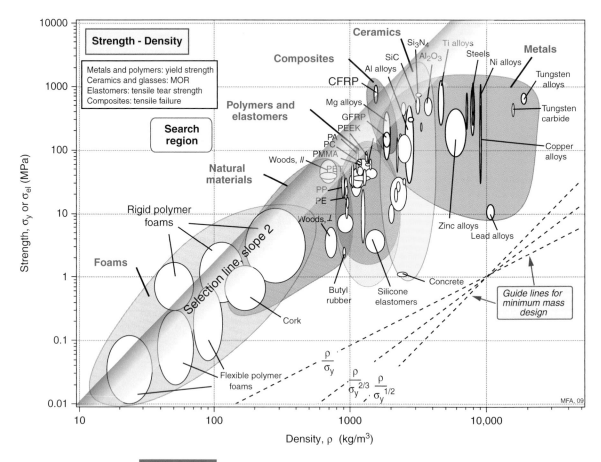

FIGURE 9.7 *A chart-based selection for a shell of prescribed strength and minimum weight. Carbon fiber reinforced polymer (CFRP) is the best choice.*

and shape. CFRP offers exceptional stiffness and strength per unit weight; making it into a doubly curved shell adds shape-stiffness, further enhancing performance.

9.5 Heating and cooling

Heating and cooling are among the most energy-gobbling, CO_2-belching things we do. Refrigerators, freezers, and air conditioners keep things cold. Central heating, ovens, and kilns keep things hot. For all these, it is the use phase of life that contributes most to energy consumption and emissions. Some, like refrigeration, central heating, or cooling, aim to hold temperatures constant over long periods of time—the fridge or building is heated or cooled once and then held like that. Others, like ovens and kilns, heat up and cool down every time they are used, zigzagging up and down in temperature over the span of a few hours. The best choice of material to minimize heat loss depends on what the use cycle looks like.

Refrigerators. To get into the topic, take a look at refrigerators. The function of a fridge is to provide a cold space. A fridge is an energy-using product (an "EuP"), and like most EuPs, it is the use phase of life that dominates energy consumption and emission release. Thus a measure of eco-excellence for a fridge is the *energy per year per cubic meter of cold space*, H_f^*, the * signifying "per cubic meter." Minimizing this is the objective.

But suppose that the fridges that are good by this criterion are expensive and the ones that are not so good are cheap. Then economically minded consumers will perceive a second measure of excellence in choosing a fridge: *the initial cost per cubic meter of cold space, C_f^**. Minimizing this becomes an objective, one that, almost certainly, conflicts with the first. Resolving the conflict needs the tradeoff methods of Chapter 8.

Figure 9.8 plots the two measures of excellence for 95 contemporary (2008) fridges. The tradeoff line is sketched (remember that it is just the convex-down envelope of the occupied space). The fridges that lie on or near it are the best choices; several are identified. They are "nondominated solutions"—offering lower energy for the same price or lower price for the same energy than any of the others.

That still leaves many, and they differ a lot. So wheel out *penalty functions*. The easiest unit of penalty is that of cost, in whatever currency you choose (here US$). We want to minimize life cost, which we take to be the sum of the initial cost and the cost of the energy used over the product's life. So, define the penalty function

$$Z^\star = C_f^\star + \alpha_e H_f^\star t \qquad (9.10)$$

where α_e, the exchange constant, is the cost of energy per kW.hr and t is the service life of the fridge in years, making $Z^\star$ the life cost of the fridge per cubic meter of cold space. The grand objective is to minimize $Z^\star$.

Take the service life to be 10 years and the cost of electrical power to be US$0.2 per kW.hr. Then the penalty function becomes

$$Z^\star = C_f^\star + 2H_f^\star \qquad (9.11)$$

or, solving for $H_f^\star$:

$$H_f^\star = \frac{1}{2}Z^\star - \frac{1}{2}C_f^\star \qquad (9.12)$$

The axes of Figure 9.8 are $H_f^\star$ and $C_f^\star$, so this equation describes a family of straight lines with a slope of $-1/2$, one for any given value of penalty $Z^\star$. Five are shown for $Z^\star$ values between $2000 and $6000. The best choices are the fridges with the lowest value of $Z^\star$—the ones where the $Z^\star$ contour is tangent to the tradeoff line.

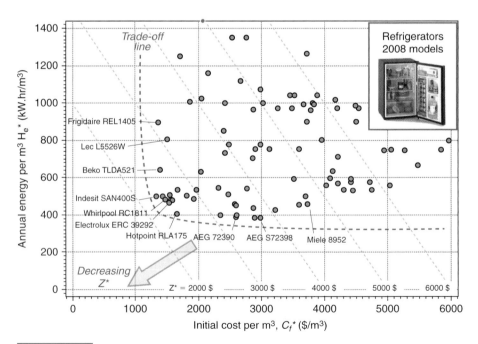

FIGURE 9.8 *A tradeoff plot for fridges with contours of the penalty function Z*. Data from 2008 advertising.*

If by some miracle the cost of energy drops by a factor of 10, the Z^* contours get 10 times steeper—almost vertical—and the best choice becomes the cheapest fridge, regardless of power consumption. If, more probably, it rises by a factor of 10, the contours become almost flat and the best choice shifts to those that use least energy, regardless of initial cost.

You could argue that this purely economic view of selection is misguided. The environment is more important than that; reducing use energy and emissions has a greater value than $0.2 per kW.hr. Fine. Then you must define what you believe it to be worth and use that for α_e instead, basing your selection on the Z^* contours that result. But if you want the rest of the world to follow your example, you must persuade them—or if you are in government, make them—use the same value. There are no absolutes in this game. Everything has to be assigned a value.

Figure 9.8 is a tradeoff plot with a sharp nose at the lower left. When the plots are like this, the optimal choice is not very sensitive to the value of the exchange constant α_e. But when the nose is more rounded, its value influences choice more strongly. We will see an example in a later case study.

So fridges use energy. Where does it go? And what choice of material would minimize it? To answer that we need a little modeling.

Modeling Thermal Loss. Creating and maintaining cold or hot space costs energy. The analysis is the same for both (though the choice of material is not). The result depends on how often the space is heated and cooled and how long it is held like that on each cycle. Take, as a generic example, the heated space, oven, or kiln sketched in Figure 9.9; we will refer to it as "the kiln." The design requirements are listed in Table 9.5.

When a kiln is fired, the internal temperature rises from ambient, T_o, to the operating temperature, T_i, where it is held for the firing time t. The

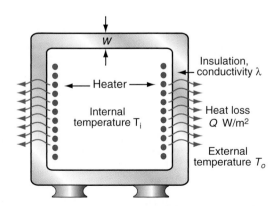

FIGURE 9.9   *A heated chamber with heat loss by conduction through the insulation.*

Table 9.5	Design requirements for kiln wall	
Function	**Thermal insulation for kiln (cyclic heating and cooling)**	
Constraint	Maximum operating temperature 1000°C (according to choice) Limit on kiln wall thickness for space reasons	
Objectives	Minimize energy consumed cyclic heating	
Free variables	Kiln wall thickness, w Choice of material	

energy consumed in one firing has two contributions. The first is the heat absorbed by the kiln wall in raising it to T_i. Per unit area, it is

$$Q_1 = C_p \rho w \left(\frac{T_i - T_o}{2} \right) \tag{9.13}$$

where C_p is the specific heat of the wall per unit mass (so $C_p \rho$ is the specific heat per unit volume) and w is the insulation wall thickness. It is minimized by choosing a wall material with a low heat capacity $C_p \rho$ and by making it as thin as possible.

The second contribution is the heat conducted out: at steady state the heat loss by conduction, Q_2, per unit area, is given by the first law of heat flow. If held for time t it is

$$Q_1 = -\lambda \frac{dT}{dx} t = \lambda \frac{(T_i - T_o)}{w} t \tag{9.14}$$

It is minimized by choosing a wall material with a low thermal conductivity λ and by making the wall as thick as possible.

The total energy consumed per unit area is the sum of these two:

$$Q = Q_1 + Q_2 = \frac{C_p \rho w \Delta T}{2} + \frac{\lambda \Delta T}{w} t \tag{9.15}$$

where $\Delta T = (T_i - T_o)$. Consider first the limits when the wall thickness w is fixed. When the heating cycle is short, the first term dominates and the best choice of material is that with the lowest volumetric heat capacity, $C_p \rho$. When instead the heating cycle is long, the second term dominates and the best choice of material is that with the smallest thermal conductivity, λ.

A wall that is too thin loses much energy by conduction but little to heat the wall itself. One that is too thick does the opposite. There is an optimum thickness, which we find by differentiating Equation 9.15 with respect to wall thickness w and equating the result to zero, giving:

$$w = \left(\frac{2\lambda t}{C_p \rho}\right)^{1/2} = (2at)^{1/2} \qquad (9.16)$$

where $a = \lambda/\rho C_p$ is the thermal diffusivity. The quantity $(2at)^{1/2}$ has dimensions of length and is a measure of the distance heat can diffuse in time t. Substituting Equation 9.16 back into Equation 9.5 to eliminate w gives:

$$Q = (\lambda C_p \rho)^{1/2} \Delta T (2t)^{1/2} \qquad (9.17)$$

This is minimized by choosing a material with the lowest value of the quantity

$$M = (\lambda C_p \rho)^{1/2} = \frac{\lambda}{a^{1/2}} \qquad (9.18)$$

Figure 9.10 shows the $\lambda - a$ chart of Chapter 8, expanded to include more materials that are good thermal insulators. All three of the criteria we have derived—minimizing $C_p\rho$, λ and $(\lambda C_p\rho)^{1/2}$—can be plotted on it; the "guidelines" show the slopes. For long heating times it is λ we want to minimize, and the best choices are the materials at the bottom of the chart: polymeric foams or, if the temperature T_i is too high for them, foamed glass, vermiculite, or carbon. But if we are free to adjust the wall thickness to the optimum value of Equation 9.16, the quantity we want to minimize is $(\lambda C_p\rho)^{1/2}$. A selection line with this slope is plotted in the figure. The best choices are the same as before, but now the performance of vermiculite, foamed glass, and foamed carbon are almost as good as that of the best polymer foams. Here the limitation of the hard-copy charts becomes apparent: there is not enough room to show a large number of specialized materials such as refractory bricks and concretes. The limitation is overcome by the computer-based methods mentioned in Chapter 8, allowing a search over a much greater number of materials.

Exactly the same analysis works for refrigerators and ovens. For workspace and housing there is an additional complication: humans have to breathe, and that means ventilation, and ventilation means that hot or cold air is pumped out of the "space" and replaced by new air that has to be heated or cooled.

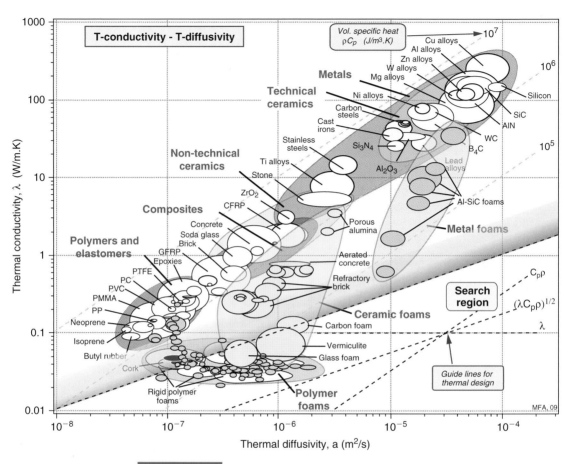

FIGURE 9.10 *The Thermal Conductivity–Thermal Diffusivity chart with contours of volumetric specific heat: the one for minimum thermal loss.*

Materials can help here too by acting as heat exchangers, extracting heat or cold from the outgoing air and using it to precondition the air coming in.

Postscript. It is not generally appreciated that, in an efficiently designed kiln, as much energy goes into heating up the kiln itself as is lost by thermal conduction to the outside environment. It is a mistake to make kiln walls too thick; a little is saved in reduced conduction loss, but more is lost in the greater heat capacity of the kiln itself.

That is the reason that foams are good; they have a low thermal conductivity *and* a low heat capacity. Centrally heated houses in which the heat is turned off at night suffer a cycle like that of the kiln. Here (because T_i is lower) the best choice is a polymeric foam, cork, or fiberglass (which has thermal properties like those of foams). But as this case study shows, turning the heat off at night doesn't save you as much as you think, because you have to supply the heat capacity of the walls in the morning.

9.6 Transport

Transport accounts for 32% of the energy we use and 34% of the emissions we generate (Figure 2.4). Cars contribute a large part of both. The primary eco-objective in car design is to *provide transport at minimum environmental impact*, which we will measure here by the CO_2 rating in grams per kilometer (g/km). The audits of Chapter 7 confirmed what we already knew: that the energy consumed during the life phase of a car exceeds that of all the other phases put together. If we are going to reduce it, we first need to know how it depends on vehicle weight and propulsion system.

Energy, carbon, and cars. The fuel consumption and CO_2 emission of cars increase with their weight. Figures 9.11 and 9.12 show the evidence for petrol, diesel, LPG, and hybrid-engine vehicles. The first shows energy plotted against mass on log scales, allowing the power-law fit for the energy consumption, H_{km}, in MJ/km and carbon emission, $CO_{2/km}$ in g/km, as a function of the vehicle mass, m, in kg, shown in the second column of Table 9.6. The second, Figure 9.12, shows the carbon rating in g/km as a function of energy per km (MJ/km) on linear scales. The two are proportional, with the constants of proportionality marked on the figure. The CO_2 rating (g/km) as a function of mass (kg) in the third column of the table is found by multiplying this by the energy/km in column 2.

From these we calculate the energy penalty associated with one kilogram of increased weight, evaluated here for a car of weight 1000 kg, by differentiating the expressions for H_m in Table 9.6. The results are listed in the last column of the table. We now have the inputs we need for modeling and selection.

But first, let us look at another aspect of car data by performing the selection that was set up in Chapter 8, in Section 8.2 and Figure 8.1: selection of a car to meet constraints on power, fuel type, and number of doors and subject to two objectives, one environmental (CO_2 rating), the other economic (cost of ownership). Figure 8.2 was a schematic of the tradeoff between the objectives. Figure 9.13a is real. It shows carbon rating and cost of ownership for 2600 cars plotted on the same axes as the schematic.[2] We are in luck: the tradeoff line has a sharp nose; the cars with the lowest CO_2 rating also have the lowest cost of ownership. The best combinations

---

[2]Data from *What Car* magazine (2005).

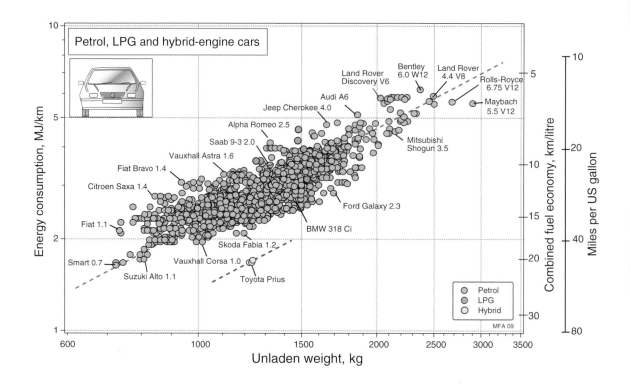

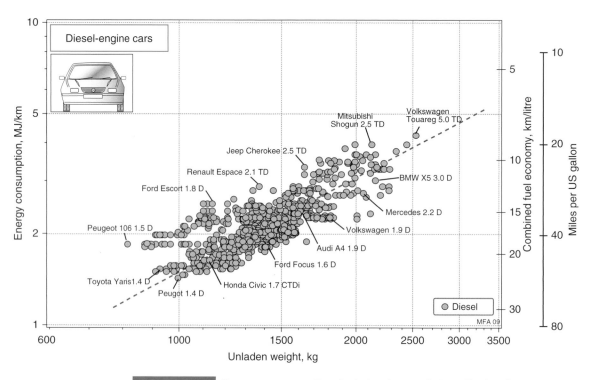

FIGURE 9.11 *Top: energy consumption of petrol engine cars; bottom: diesel-engine cars.*

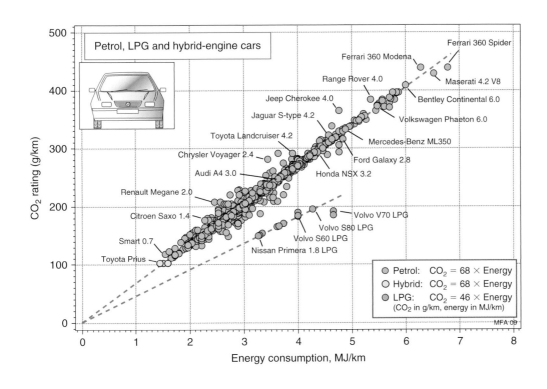

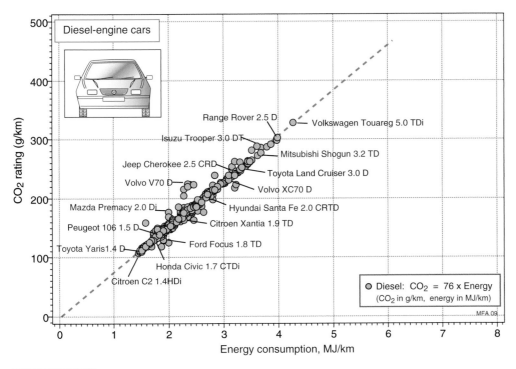

FIGURE 9.12 *Top: CO_2 emission of petrol engine cars; bottom: diesel-engine cars.*

Table 9.6	The energy and CO_2 rating of cars as a function of their mass		
Fuel Type	**Energy per km: mass (H_{km} in MJ/km, m in kg)**	**CO_2 per km: mass ($CO_{2/km}$ in g/km, m in kg)**	**dH_{km}/dm MJ/km.kg ($m = 1000$ kg)**
Petrol power	$H_{km} \approx 3.7 \times 10^{-3} m^{0.93}$	$CO_{2/km} \approx 0.25 m^{0.93}$	2.1×10^{-3}
Diesel power	$H_{km} \approx 2.8 \times 10^{-3} m^{0.93}$	$CO_{2/km} \approx 0.21 m^{0.93}$	1.6×10^{-3}
LPG power	$H_{km} \approx 3.7 \times 10^{-3} m^{0.93}$	$CO_{2/km} \approx 0.17 m^{0.93}$	2.2×10^{-3}
Hybrid power	$H_{km} \approx 2.3 \times 10^{-3} m^{0.93}$	$CO_{2/km} \approx 0.16 m^{0.93}$	1.3×10^{-3}

are the cars that lie nearest to the tradeoff line, at the lower left of the plot. Not surprisingly, they are all very small.

That is the picture before the constraints are applied. If we now screen out all models that are not gas powered, have less than 150 hp, and fewer than four doors, the tradeoff plot looks like Figure 9.13b. Again, the tradeoff line has a sharp nose. The cars that lie there have the best combination of cost and carbon and meet all the constraints; we don't need a penalty function to find them. The Honda Civic 2.0 is a winner, but the Toyota Corolla 1.8 and the Renault Laguna 2.0 lie close. It is worth exploring these in more depth. What are the service intervals? How close is the nearest service center? What do consumer magazines say about them? It is this documentation that allows a final choice to be reached.

Not all tradeoffs are so straightforward. Performance, to many car owners, is a matter of importance. For someone who wants to be responsible about carbon yet drive a high-performance vehicle, the compromise is more difficult: high performance means high carbon. Figure 9.14 shows the tradeoff, using acceleration, measured by the time from 0 to 60 mph, as one objective (the shorter the time, the greater the performance) and carbon rating as the second. Now the tradeoff line is a broad curve. As before, the best choices are those that lie on or near this line; all others (and there are many) can be rejected. To get further, it is necessary to assign relative values to performance and carbon rating. If the first is the most highly valued, it is the cars at the upper left that become the prime candidates. If it is carbon, it is those at the lower right. The compromise is a harsh one; any choice with low carbon has poor performance.

Modeling: where Does the Energy Go? Energy is dissipated in transport in three ways: as the energy needed to accelerate the vehicle up to its cruising

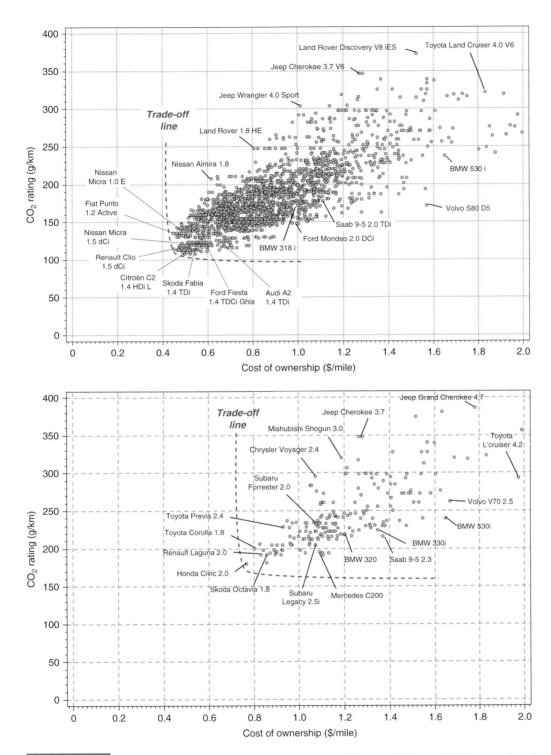

FIGURE 9.13 *The tradeoff between CO_2 rating and cost of ownership for cars (2005 models). The constraints of four doors, gas fuel, and 150+ hp have not yet been applied. Bottom: the same tradeoff after applying the constraints. The figure identifies the cars that meet the constraints and minimize the objectives.*

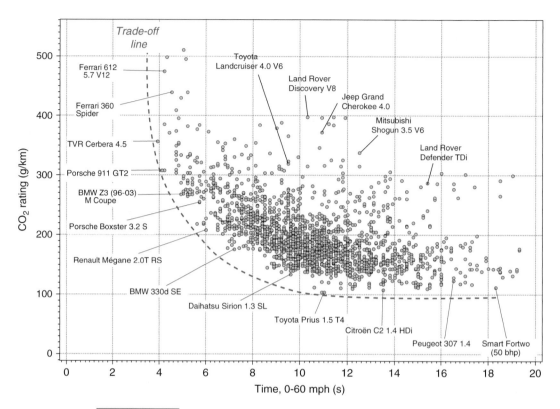

FIGURE 9.14 *A tradeoff plot performance, measured by time to accelerate from 0 to 60 mph (100 km/hour), against CO$_2$ rating for 1700 cars.*

speed, giving it kinetic energy that is lost on braking; as drag exerted by the air or water through which it is passing; and as rolling friction in bearings and the contact between wheels and road or track. Imagine (following MacKay, 2008) that a vehicle with mass m accelerates to a cruising velocity v acquiring kinetic energy:

$$E_{ke} = \frac{1}{2}mv^2. \tag{9.19}$$

It continues over a distance d for a time d/v before stopping, losing this kinetic energy as heat in the brakes, thus dissipating energy per unit time (*power*) of

$$\frac{dE_{ke}}{dt} \approx \frac{E_{ke}}{d/v} = \frac{1}{2}\frac{mv^3}{d}. \tag{9.20}$$

while cruising the vehicle drags behind it a column of air with a cross-section proportional to its frontal area A. The column created in time t has a volume $c_d A v t$ where c_d is the drag coefficient, typically about 0.3 for a car, 0.4 for a bus or truck. This column has a mass $\rho_{air} c_d A v t$ and it moves with velocity v, so the kinetic energy imparted to the air is

$$E_{drag} = \frac{1}{2} m_{air} \, v^2 = \frac{1}{2} \rho_{air} c_d A \, v^3 t \qquad (9.21)$$

where ρ_{air} is the density of air. The drag is the rate of change of this energy,

$$\frac{dE_{drag}}{dt} = \frac{1}{2} \rho_{air} c_d A \, v^3. \qquad (9.22)$$

If this cycle is repeated over and over again, the power dissipated is the sum of these two:

$$Power = \frac{1}{2}\left(\frac{m}{d} + \rho_{air} \, c_d \, A\right) v^3 \qquad (9.23)$$

Rolling resistance adds another small term that is proportional to the mass, which we will ignore.

The first term in the brackets is proportional to the mass of the vehicle and the distance it moves between stops. The second depends only on the frontal area and the drag coefficient. Thus for short haul, stop-and-go, or city driving, the way to save fuel is to make the vehicle as light as possible. For long-haul, steady cruising, or motorway driving, mass is less important and minimizing drag (meaning frontal area A and drag coefficient c_d) is key. The data for average energy consumption of Table 9.6 shows a near-linear dependence of energy per km on mass, meaning that, in normal use, it is the first term that dominates. Thus design to minimize the energy and CO_2 of vehicle use must focus on material selection to minimize mass.

Selection Making cars lighter means replacing heavy steel and cast iron components by those made of lighter materials: light alloys based on aluminum, magnesium, or titanium and composites reinforced by glass or carbon fibers. All these materials have greater embodied energy per kg than steel, introducing a new sort of tradeoff: that between the competing energy demands of different life phases. There is a net saving of energy and CO_2 only if that saved by weight reduction exceeds that invested as extra embodied energy. Figure 9.15 illustrates the problem. It shows what happens when an existing material (which we take to be steel for this example) for a vehicle

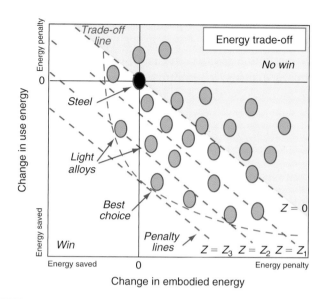

FIGURE 9.15 *The use energy change and the embodied energy change when a steel component is replaced by one made of a light alloy. There is net energy saving only if the sum of the two is negative. The diagonal lines are contours of constant (negative) sum.*

component is replaced with a substitute. The horizontal axis plots the change in embodied energy, ΔH_{emb}. The vertical one plots the change in use energy, ΔH_{use}. The black circle at (0,0) is the steel; the other circles, enclosed by a tradeoff line, represent substitutes. The diagonal contours show the penalty function, which, in this instance, is particularly simple:

$$Z = \Delta H_{emb} + \Delta H_{use}$$

The best choice of material is that with the lowest (most negative) value of Z. Any substitute that lies on the contour of $Z = 0$ passing through steel offers no reduction or increase in life energy. Those lying in the blue "No win" zone give an increase. Those in the white "Win" zone offer a saving. The best choice of all is that nearest the point that a penalty contour is tangent to the tradeoff line; it is indicated on the figure.

All very straightforward. Well, not quite. Material substitution in a component that performs a mechanical function requires that the component be rescaled to have the same stiffness or strength (whichever is design limiting) as the original. In making the substitution, the mass of the component changes both because the density of the new material differs from that of the old and because the scaling changes the volume of material that is used.

The scaling rules are known; they are given by material indices developed in Chapter 8. They allow the change in *mass* Δm and in *embodied energy* ΔH_{emb} of the component resulting from the change of material to be calculated. Multiplying Δm by the energy per kg.km from Table 9.6 and the distance traveled over life, which we will take as 200,000 km, gives the change in *use-phase energy* ΔH_{use} resulting from substitution.

Consider, then, the replacement of the pressed steel bumper set of a car by one made of a lighter material. The function and weight scaling of the bumper were described earlier: it is a beam of given bending strength with a mass, for a given bending strength, that scales as $\rho/\sigma_y^{2/3}$, where ρ is the density and σ_y is the yield strength of the material of which it is made. The weight change of the bumper set, on replacing one made of steel (subscript 0) with one made of a lighter material (subscript 1), is thus:

$$\Delta m = B \left(\frac{\rho_1}{\sigma_{y,1}^{2/3}} - \frac{\rho_o}{\sigma_{y,0}^{2/3}} \right) \qquad (9.24)$$

where B is a constant. If the mass m_o of the steel bumper set is 20 kg, then

$$m_o = B \frac{\rho_o}{\sigma_{y,0}^{2/3}} = 20 \, \text{kg} \qquad (9.25)$$

defining B. Substituting this value for B into the previous equation gives

$$\Delta m = 20 \left(\frac{\rho_1}{\sigma_{y,1}^{2/3}} \left[\frac{\sigma_{y,0}^{2/3}}{\rho_o} \right] - 1 \right) \qquad (9.26)$$

The property group in the square brackets is that for the steel of the original bumper; we will use data for an AISI 1022 rolled steel with 0.18% C 0.7% Mn, a yield strength of 295 MPa and a density of 7,900 kg/m³, giving the quantity in square brackets the value, based on these units, of 5.6×10^{-3}. From Table 9.6, a gasoline-engine vehicle weighing 1000 kg requires 2.1×10^{-3} MJ/km.kg. Thus the change in use energy, ΔH_{use}, found by multiplying this by Δm and the distance traveled over a life of 200,000 km, is

$$\Delta H_{use} = 8.4 \times 10^3 \left(5.6 \times 10^{-3} \left[\frac{\rho_1}{\sigma_{y,1}^{2/3}} \right] - 1 \right) \text{MJ} \qquad (9.27)$$

The change in embodied energy ΔH_{emb} is found in a similar way. The embodied energy of the initial steel bumper set is:

$$H_{emb} = mH_{mo} = CH_{mo}\frac{\rho_o}{\sigma_{y,0}^{2/3}} = 20 \times 33\,\text{MJ} \qquad (9.28)$$

where $m = 20\,kg$ is the mass and $H_{mo} = 33\,MJ/kg$ the embodied energy of the original steel bumper set. The constant C is defined by this equation. The change on switching to a new material is then:

$$\Delta H_{emb} = C\left(\frac{H_{m1}\rho_1}{\sigma_{y,1}^{2/3}} - \frac{H_{mo}\rho_o}{\sigma_{y,0}^{2/3}}\right) = 660\left[\frac{H_{m1}\rho_1}{\sigma_{y,1}^{2/3}}\left[\frac{\sigma_{y,0}^{2/3}}{H_{mo}\rho_o}\right] - 1\right] \qquad (9.29)$$

As before, the property group in the square brackets is that for the steel of the original bumper. Its value, using steel data given previously, is 1.7×10^{-4}, giving the final expression for change in embodied energy:

$$\Delta H_{emb} = 660\left(1.7 \times 10^{-4}\frac{H_{m1}\rho_1}{\sigma_{y,1}^{2/3}} - 1\right)\text{MJ} \qquad (9.30)$$

The tradeoff between ΔH_{use} and ΔH_{emb} is plotted in Figure 9.16. It explores the total energy saving (or lack of it) when the carbon steel bumper set is replaced by one made of a low alloy steel, a 6000 series aluminum alloy, a wrought magnesium alloy, or a composite. Steel lies at the origin (0,0). The other bubbles, labeled, show the changes in ΔH_{use} and ΔH_{emb} calculated from Equations 9.27 and 9.30. The "No win" area is shaded. For titanium alloys, the energy saving is slight. For magnesium and aluminum alloys, it is considerable. The greatest saving is made possible using composites. The tradeoff line is sketched in. The penalty contours show the total energy saved over 200,000 km. The contour that is tangent to the tradeoff line has (by interpolation) a value of about 7 GJ. It identifies the best choice: here, the glass-epoxy laminate.

Before a final decision is reached, it is helpful to have a feeling for the likely change in cost. The change in use cost ΔC_{use} is the use energy change ΔH_{use} (Equation 9.27) multiplied by the price of energy; gasoline at $0.8/liter ($3 per U.S. gallon) gives an energy price of approximately 0.025$/MJ. Thus:

$$\Delta C_{use} = 210\left(5.6 \times 10^{-3}\frac{\rho_1}{\sigma_{y,1}^{2/3}} - 1\right)\$ \qquad (9.31)$$

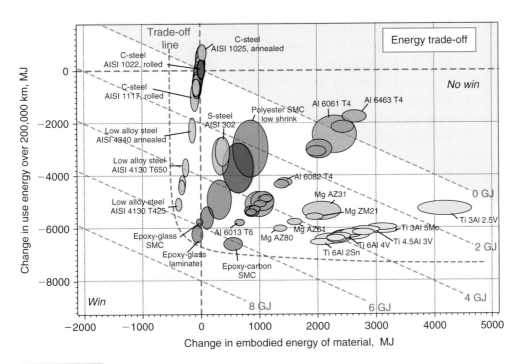

FIGURE 9.16 *The tradeoff between embodied energy and use energy.*

The change in material cost ΔC_{mat} might seem (in parallel with that of embodied energy) to be the change in the product of component mass m and the price per unit mass C_m, and we shall use this:

$$\Delta C_{mat} = C\left[\frac{C_{m1}\rho_1}{\sigma_{y,l}^{2/3}} - \frac{C_{mo}\rho_o}{\sigma_{y,0}^{2/3}}\right] = 20C_{mo}\left[\frac{C_{m1}\rho_1}{\sigma_{y,l}^{2/3}}\frac{\sigma_{y,0}^{2/3}}{C_{mo}\rho_o} - 1\right] \quad (9.32)$$

Taking the cost C_{mo} of steel to be \$0.8/kg makes the value of the steel property group in the square brackets equal to 7.0×10^{-3} and the equation becomes:

$$\Delta C_{mat} = 16\left[7.0 \times 10^{-3}\left(\frac{C_{m1}\rho_1}{\sigma_{y,l}^{2/3}}\right) - 1\right]\$ \quad (9.33)$$

But this ignores manufacture. Cost has other contributions; at least half the cost of a component is that of manufacturing. The dominant feature in this is *time*. If forming, joining, and finishing with the new material are slower than with the old, there is an additional cost penalty. So the material cost of Equation 9.13 must be regarded as approximate only.

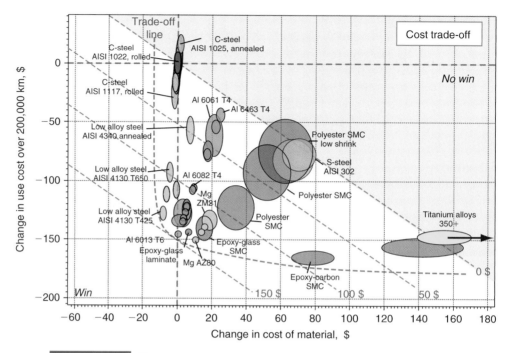

FIGURE 9.17   *The tradeoff between material cost and use cost.*

With this simplifying approximation, the cost tradeoff (following the pattern used for energy) appears as in Figure 9.17. Titanium alloys lie well inside the "No win" zone. Low alloy steel, aluminum alloys, and magnesium alloys are the most attractive from an economic point of view, though epoxy glass is almost as good. The most striking result is that the total sum saved is so small.

Postscript. We have calculated the primary mass saving that material substitution brings. In reality the saving is—or can be—even larger because the lighter vehicle can get away with less heavy suspension, lighter tires, and less powerful brakes, allowing a secondary weight saving. In current practice, however, aluminum cars are not much lighter than the steel ones they replace, because manufacturers tend to load them with more extras (such as air conditioning as standard), adding mass back on.

9.7 Summary and conclusion

Rational selection of materials to meet environmental objectives starts by identifying the phase of product life that causes greatest concern: production, manufacture, transport, use, or disposal. Dealing with all these requires data not only for the obvious eco-attributes (energy, emissions, toxicity, ability to be recycled, and the like) but also data for mechanical, thermal, electrical, and chemical properties. Thus if material production is the phase of concern, selection is based on minimizing embodied energy or the associated emissions (CO_2 release, for example). But if it is the use phase that is of concern, selection is based instead on light weight or excellence as a thermal insulator or electrical conductor. These define the objective; the idea is to minimize this while meeting all the other constraints of the design: adequate stiffness, strength, durability, and the like.

Almost always there is more than one objective, and almost always they conflict. They arise in more than one way. One is the obvious conflict between eco-objectives and cost, illustrated both in Chapter 8 and here, for which trade-off methods offer a way forward, provided a value can be assigned to the eco-objective, something that is not always easy. Another is the conflict between the energy demands and emissions of different phases of life: the conflict between increased embodied energy of material and reduced energy of use, for example. Tradeoff methods work particularly well for this type of problem because both energies can be quantified. The case studies of this chapter illustrate how such problems are tackled. The exercises of Section 9.9 present more.

9.8 Further reading

Ashby, M.F. (2005), "Materials selection in mechanical design", 3rd ed, Butterworth Heinemann. Chapter 4, ISBN 0-7506-6168-2. (*A text that develops the ideas presented here in more depth, including the derivation of material indices, a discussion of shape factors, and a catalog of simple solutions to standard problems.*)

Ashby, M.F., Shercliff, H.R. and Cebon, D. (2007), "Materials: engineering, science, processing and design", Butterworth Heinemann. ISBN-13: 978-0-7506-8391-3. (*An elementary text introducing materials through material property charts and developing the selection methods through case studies.*)

MacKay, D.J.C. (2008), "Sustainable energy – without the hot air, Department of Physics", UIT Press, Cambridge UK. ISBN 978-0-9544529-3-3. (*MacKay brings common sense into the discussion of energy use.*)

Caceres, C.H. (2007), Economical and environmental factors in light alloys automotive applications, Metallurgical and Materials Transactions. (*An analysis of the cost-mass tradeoff for cars.*)

Solutions to standard problems: (a) stiffness and strength

Calladine, C.R. (1983), "Theory of shell structures", Cambridge University Press. ISBN 0-521-36945-2

Young, W.C. (1989), "Roark's formulas for stress and strain", 6th ed, McGraw-Hill. ISBN 0-07-072541-1. (*A "yellow pages" for results for calculations of stress and strain in loaded components.*)

Solutions to standard problems: (b) heat flow

Hollman, J.P. (1981), "Heat transfer", 5th ed, McGraw Hill. ISBN 0-07-029618-9

Carslaw, H.S. and Jaeger, J.C. (1959), "Conduction of heat in solids", 2nd ed, Oxford University Press. ISBN 0-19-853303-9

9.9 Exercises

E.9.1. The materials of the drink containers of Figure 9.1 are recycled to different degrees. How does the ranking of Table 9.2 change if the contribution of recycling is included? To do so, multiply the energy per liter in the last column of the table by the factor

$$1 - f_{rc}\left(1 - \frac{H_{rc}}{H_m}\right)$$

where f_{rc} is the recycle fraction in current supply, H_m is the embodied energy for primary material production, and H_{rc} is that for recycling of the material. You will find data for all three attributes in the data sheets of Chapter 12.

E.9.2. Derive the correction factor to allow for recycle content cited in Exercise E.9.1.

E.9.3. Repeat the analysis of Table 9.2 and Section 9.2, applying it to industrial fluid containers of the sort you find in automobile supply stores: those for antifreeze, oil, cleaning fluids, and gasoline. Weigh them, record their volume, identify the material of which they are made (recycle mark? magnetic? non-magnetic?), retrieve their embodied energies from the data sheets of Chapter 12, and rank them by embodied energy per unit volume of fluid contained.

E.9.4. Use the indices for the crash barriers (Equations 9.1 and 9.2) with the charts for strength and density (Figure 8.12) and strength and embodied energy (Figure 8.14) to select materials for each of the barriers. Position your selection line to include one metal for each. Reject ceramics and glass on the grounds of brittleness. List what you find for each barrier.

E.9.5. Complete the selection of materials for light, stiff shells of Section 9.5 by plotting the stiffness index

$$M_1 = \frac{\rho}{E^{1/2}}$$

onto a copy of the modulus-density chart of Figure 8.11. Reject ceramics and glass on the grounds of brittleness and foams on the grounds that the shell would have to be very thick. Which materials do you find? Which of these would be practical for a real shell?

E.9.6. In a faraway land, refrigerators cost the same as they do here, but electrical energy costs 10 times more than here—that is, it costs $2/kW.hr. Make a copy of the tradeoff plot for fridges (Figure 9.8) and plot a new set of penalty lines onto it, using this value for the exchange constant, α_e. If you had to choose just one fridge, which would it be?

E.9.7. You are asked to design a large heated workspace in a cold climate, making it as ecofriendly as possible by using straw-bale insulation. Straw, when compressed, has a density of 600 kg/m^3, a specific heat capacity of 1670 J/kg.K, and a thermal conductivity of 0.09 W/m.K. The space will be heated during the day (12 hours) but not at night. What is the optimum thickness of straw to minimize the energy loss?

E.9.8. The makers of a small electric car want to make bumpers out of a molded thermoplastic. Which index is the one to guide this selection? Plot it on the appropriate chart from the set shown in Figures 8.11–8.14 and make a selection.

E.9.9. Car bumpers used to be made of steel. Most cars now have extruded aluminium or glass-reinforced polymer bumpers. Both materials have a much higher embodied energy than steel. Take the mass of a steel bumper set to be 20 kg and that of an aluminum one to be 14 kg. Find an equation for the energy consumption in MJ/km as a function of weight for petrol engine cars using the data plotted in Figure 9.11 of the text.

- Work out how much energy is saved by changing the bumper set of a 1500 kg car from steel to aluminium over an assumed life of 200,000 km.

- Calculate whether the switch from steel to aluminum has saved energy over life. You will find the embodied energies of steel and aluminum in the datasheets of Chapter 12. Ignore the differences in energy in manufacturing the two bumpers; it is small.

- The switch from steel to aluminum increases the price of the car by $60. Using current pump prices for gasoline, work out whether, over the assumed life, it is cheaper to have the aluminum bumper or the steel one.

Exercises Using the CES Edu Software

E.9.10. Refine the selection for shells using Level 3 of the CES software. Make a chart with the two indices

$$M_1 = \frac{\rho}{E^{1/2}} \text{ and } M_2 = \frac{\rho}{\sigma_y^{1/2}}$$

as axes, using the Advanced facility to make the combination of properties. Then add a Tree stage, selecting only metals, polymers, and composites and natural materials. Which ones emerge as the best choice? Why?

E.9.11. Tackle the crash-barrier case study using CES Level 2 following the requirements set out in Table 9.3. Use a Limit stage to apply the constraints on fracture toughness $K_{lc} \geq 18\,MPa.m$ and the requirement of recyclability. Then make a chart with density ρ on the x axis and yield strength σ_y on the y axis, and apply a selection line with the appropriate slope to represent the index for the mobile barrier:

$$M_2 = \frac{\rho}{\sigma_y^{2/3}}$$

List what you find to be the best candidates. Then replace Level 2 with Level 3 data and explore what you find.

E.9.12. Repeat the procedure of Exercise E.9.11, but this time make a chart using CES Level 2 on which the index for the static barrier

$$M_1 = \frac{H_m \rho}{\sigma_y^{2/3}}$$

can be plotted. You will need the Advanced facility to make the product $H_m \rho$. List what you find to be the best candidates. Then, as before, dump in Level 3 data and explore what you find.

Sustainability: living on renewables

10.1 Introduction and synopsis

Sustainability is one of those words that has come to mean whatever the speaker wants it to mean. To large corporations it means staying in business and continuing to grow; sustainability to an oil company, for instance, means adequate reserves of oil in the short term and a stake in the energy technology that replaces it when the time comes. To those who think on a broader scale, sustainability means using technology to decouple gross domestic product (GDP)[1] from environmental damage (carbon emissions,

---

Renewable and nonrenewable energy. (Image of wind turbine courtesy of Leica Geosystems, Switzerland. Image of oil rig courtesy of the U.S. National Park Service.)

---

[1] The annual GDP is an indicator used to gauge the health of a country's economy. It is the total value of all goods and services produced over a given year. Think of it as the size of the economy. GDP per capita is a measure of average individual income.

for instance), allowing the first to grow while reducing the second, using market forces such as carbon trading as a mechanism. There are people, however, who have no faith in this approach, seeing the free market as the *cause* of the problem, not the *solution*; it is the relentless drive for growth that threatens the planet. Sustainability, to them, means capping or reducing capital GDP. And there are all shades of viewpoints in between.

Here we examine the meaning of sustainability. The view you take of it depends on scale—on the time frame and spatial scope of the examination. We start with these and then explore sustainable and quasi-sustainable sources of energy and of materials.

10.2 The concept of sustainable development

As engineers and scientists it is natural that we should want to tackle problems at a level to which we can bring our skills to bear. The methods described in the previous chapters are examples. They address immediate problems, ones that are already evident and identified. But they do little to tackle the deeper problem: that of long-term *sustainability*. Figure 10.1 introduces the concept. The horizontal axis describes the time scale, ranging from that of the life of a product to that of the span of a civilization. The vertical axis describes the spatial scale, again ranging from that of the

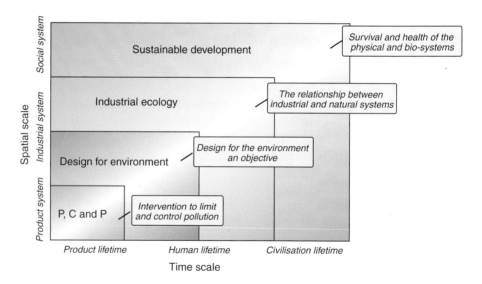

FIGURE 10.1 *Approaches, differing in spatial and temporal scale of thinking, about the industrialization and the natural ecosystem. (Adapted from Coulter et al., 1995)*

product to that of society as a whole. It has four nested boxes, expanding outward in conceptual scale, each representing an approach to thinking about the environment.

The least ambitious of these—the smallest box—is that of *pollution control and prevention* (PC and P). This is intervention on the scale and lifetime of a single product and is frequently a cleanup measure. Taking transport as an example, it is the addition of catalytic converters to cars, a step to mitigate an identified problem with an existing product or system.

The next box is that describing *design for the environment* (DFE)—the techniques discussed in Chapters 7 through 10 of this book. Here the time and spatial scales include the entire design process; the strategy is to foresee and minimize the effects of product families at the design stage, balancing them against the conflicting objectives of performance, reliability, quality, and cost. Retaining the example of the car, it is to redesign the vehicle, giving emphasis to the objectives of minimizing emissions by reducing weight and adopting an alternative propulsion system—hybrid, perhaps, or electric.

The third box, that of *industrial ecology*, derives from the precept that we must see human activities as part of the global ecosystem. Here the idea is that a study of the processes and balances that have evolved in nature might suggest ways to reconcile the imbalance between the industrial and the natural systems, an idea known as the *ecological metaphor*. We return to this idea in a moment as a way of structuring thinking about the last box, that of *sustainable development*.

10.3 The ecological metaphor

The ecological metaphor has its genesis in the observation that natural and industrial systems have certain features in common. Consider three:

1. Both the natural and the industrial systems transform resources— materials and energy—that of nature through growth, that of industry through manufacture. The plant kingdom captures energy from the sun, carbon dioxide from the atmosphere, and minerals from the Earth to create carbohydrates; the animal kingdom derives its energy and essential minerals from those of plants or from each other. The industrial system, by contrast, acquires most of its energy from fossil fuels and its raw materials from those that occur naturally in the Earth's crust, in the oceans, and in the natural world.

2. Both systems generate waste—the natural system through metabolism and death, the industrial system through the emissions

of manufacture and through the obsolescence and finite life of the products it produces. ⌈The difference is that the waste of nature is recycled with 100% efficiency,⌋ allowing a steady state, drawing on renewable energy (sunlight) to do so. The waste of industry is recycled much less effectively, doing so with nonrenewable energy (fossil fuels).

3. Both the natural and the industrial systems exist within the ecosphere, which provides the raw materials and other primary resources, acts as a reservoir for waste, absorbing and, in nature, recycling it, and providing the essential environment for life, meaning fresh water, a breathable atmosphere, protection from UV radiation, and more. The natural system manages, for long periods, to live in balance with the ecosphere. Our present industrial system, it appears, does not. Are there lessons to be learned about managing industrial systems from the balances that have evolved in nature? Can nature give guidance or at least provide an ideal?

The most important elements in living things are carbon, nitrogen, hydrogen, and oxygen; they make up the carbohydrates, fats, and proteins on which life depends. The only way that plants and animals can continue to take in and use nutrients containing these elements is if they are constantly recycled around the ecosystem for reuse. The most important of these circular paths are the *carbon cycle*, the *nitrogen cycle,* and the *hydrological (water) cycle*. Subsystems have evolved that provide the links in the cycles and do so at rates that match those of the subsystems with which they interface. Figure 10.2 is a sketch of one of these, the carbon cycle. Carbon dioxide in the atmosphere is captured by green plants and algae on land and by phytoplankton and other members of the aquatic biomass in water. Fungal and bacterial action enables decomposition of plants and animals when they die, returning much of the carbon to the atmosphere but also sequestering some as carbon-rich deposits (peat, gas, oil, coal) and, in the oceans, as limestone, $CaCO_3$.

The elements important for manmade products, by contrast, are far more numerous; they include most of the periodic table. Carbon is one. As in nature, the products that use it (in the form of coal, oil, or gas) return most of it to the atmosphere, but the natural subsystems that recycle carbon have not evolved to provide matching rates (Figure 10.3). The problem is more acute with other elements—the heavy metals, for example—where no natural subsystems exist to provide recycling. When rates don't match, stuff piles up somewhere. Focusing on carbon again, this imbalance is evident in the steep rise of atmospheric carbon since 1850 (Figure 3.8). Burning fossil fuels and calcining limestone for cement generate large masses of CO_2. Reduced forestation, rising water temperatures,

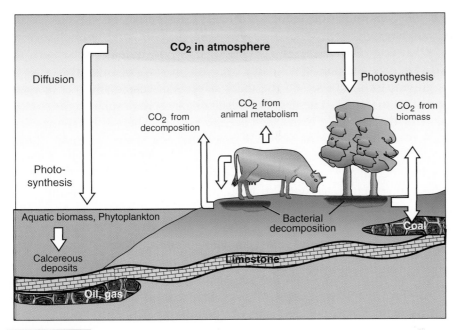

FIGURE 10.2 *The carbon cycle in nature, showing some of the many subsystems that have evolved to transform resources. They do so in such a way as to give balance and long-term stability.*

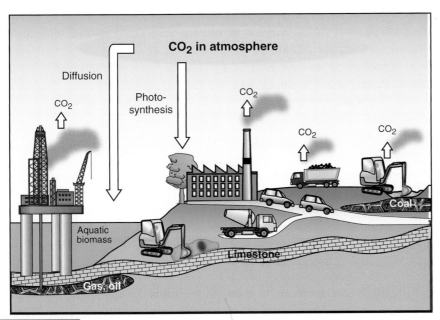

FIGURE 10.3 *The additional burden placed on the carbon cycle by large-scale industrialization. The subsystems that evolved to balance the cycle still exist but work at rates that do not begin to replace the resources that are consumed.*

and soil contamination reduce the rate of absorption on the part of the biosubsystems.

What do we learn? Figure 10.4 summarizes the differences between the two systems—the one, sustainable over long periods of time; the other, in its present form, not so. Of the many aspects of sustainability, two relate directly to materials. The first is the lack of appropriate subsystems to close many of the recycling paths, and second is that, where subsystems exist, there is an imbalance of rates. At base, however, there is a more fundamental difference: it has to do with metrics of well-being. That of nature is achieving balance, such that the system is in *equilibrium*. That of the industrial system is *growth*: an economy that is growing is healthy, one that is static is sick. Economic growth, our metric of the well-being of a business, nations, or society as a whole, carries with it the need for ever-increasing consumption of materials and energy (Chapter 2) and of waste creation (Chapter 4). The present system of industrial production has been likened to an organism that ingests resources, produces goods, and expels waste. The characteristic that makes this organism devastating for the environment is its insatiable appetite; the faster it ingests resources, the greater is the output of products and the better is its health, even though this does not coincide with that of the biosphere.[2] The comparison, then, highlights the ideal: an industrial system in which the consumption of materials and energy and the production of waste are minimized, and the discarded material from one process becomes the raw material for another, ultimately closing the loop.

The natural ecosystem	The industrial system
▪ *Uses few elements* (mainly C, N, O and H)	▪ *Uses most of the periodic table*
▪ *Is cyclic* – materials circulate and transform continuously	▪ *Is linear* – transforms materials into products and waste
▪ *Subsystems* have evolved that use "waste" as a resource	▪ *Lack of subsystems* that use "waste" as a resource
▪ *Closed loop* – no waste: each subsystem provides sustenance for others	▪ *Open loop* – waste destructive of sources on which it depends
▪ *Indicator of well-being* – equilibrium	▪ *Indicator of well-being* – growth

FIGURE 10.4 *The comparison of the natural and the industrial ecosystems.*

[2] Frosch and Gallopoulos, 1989; Regge and Pallante, 1996, cited by Guidice et al., 2006.

Are natural systems really at equilibrium? Over long periods of time, yes. The forces for change are minimal, allowing optimization at an ever more refined and detailed level. It leads to an interdependence on a scale that even now we do not fully grasp but which manmade activities too frequently disturb. However, on a geological time scale, there have been disruptions of the natural system on grand scales. Most derived from sudden climate change. Are there lessons in the way the natural system then adapted? When dinosaurs (reptiles) succumbed to one of these, some small, furry mammal—a mouse, perhaps—because of habitat, metabolism, diet— saw its opportunity, survived, evolved, and multiplied. Where, in the technical world of today, are the post-industrial mice, and what do they look like?

We don't know. But there is a message here. The life raft—the Noah's Ark, so to speak, of the natural world, allowing continuity in times of change—lies in its diversity: new mice, ready and waiting. The nuclei of the new system existed in the old and could emerge and grow when circumstances change. Without knowing what we will need to do, can we create a society, a scientific society, with sufficient diversity that, painful though change will be, the nuclei of the next phase preexist in it?

10.4 Sustainable energy

With enough cheap, nonpolluting energy, all things are possible. The average power needs of a developed country today range from 4–14 kW per person, which translates into an energy consumption of 120–450 GJ/year per person.[3] This energy is used in the ways listed in Table 10.1. The transport sector is more dependent than the others on fossil fuels; there are, at

Table 10.1 Energy consumption by sector	
Sector	**Proportion (%)**
Transport	32
Domestic	29
Industry	35
Other (food)	4

[3] 1 kW.hr = 3.6 MJ; thus 1 GJ/year per person = 3,600 kW.hr/year per person. 1 kW.hr per day = 0.042 kW.

present, no viable alternatives to gasoline, diesel, and kerosene for vehicle and aircraft propulsion.

There are, ultimately, only four sources of energy: the sun, which drives the winds, wave, hydro, and photochemical phenomena; the moon, which drives the tides; radioactive decay of unstable elements giving geothermal heat from the Earth's crust or, when concentrated, nuclear power; and hydrocarbon fuels, the sun's energy in fossilized form. All, ultimately, are finite resources, but the time scale for the exhaustion of the first three— solar, lunar, and geothermal—is so large that it is safe to regard them as infinite. And it is these three that offer the possibility of power generation without atmospheric pollution. So let us examine each in turn.

We have MacKay (2008) to thank for introducing some simple physics and a great deal of common sense into the discussion of renewable energy. The paragraphs that follow derive from his analysis.

Wind. The problem with wind power, like that of most other renewable energy sources, is the low *power density*, that is, power per unit area that can be harvested. On land, it averages $2\,W/m^2$; offshore it is larger, about $3\,W/m^2$. The average land area per person averaged across a country with a population density like that of the United Kingdom is about $3500\,m^2$. That means that if the *entire country* were packed with the maximum possible number of wind turbines, it would generate just $7\,kW$ per person. Placing them offshore helps solve the overcrowding problem, but maintenance costs are high.

Solar. Solar energy density depends where you are on the Earth's surface. In a temperate-zone country it can be as high as $50\,W/m^2$. It can be captured in a number of ways. *Solar-thermal* systems use the sun's energy to heat water with an efficiency of perhaps 50%, capturing $25\,W/m^2$. Hot water is low-grade energy, not readily transmitted or converted to other energy forms, so it can help with home heating, but not much else. *Photovoltaics* convert photons to electrical power with an efficiency of between 10% (cheap solar cells) and 20% (very expensive ones). The solar cells would have to be cheap if used to cover large areas, capturing about $5\,W/m^2$, again requiring a vast area to capture enough energy to contribute much to daily needs. We don't need water pipes and solar cells to capture the sun's energy, of course. Plants can do it for us, and then we can burn them or ferment them or even eat them to make use of the energy they stored (energy from *biomass*). But the capture efficiency of even the most efficient plants is only 1%, and commercial agriculture processing consumes energy in tractors, fermentation vats, and distillation equipment, reducing the effective efficiency to perhaps 0.5%.

Hydro. The power that can be extracted from hydroelectric schemes is limited by altitude and rainfall. In mountainous areas where it is practical to catch and store the rain, releasing the water through a height drop to a turbine, it is possible to capture $0.2\,W/m^2$—not a lot but enough to make significant contributions to power generation in Switzerland and to provide Norway with almost all its electricity.

Waves. Energy can be captured from waves by placing something in their path—a fixed barrier with a turbine driven by the water whooshing in and out, for instance. Thus waves carry an energy per unit length rather than an energy per unit area, and it is large—as much as $40\,kW$ per meter. Capturing it, however, is not easy; it is unlikely that any wave machine would trap more than a third of this power. Few countries have long coastlines (some have none), but for those that do, wave power is an option. But once again the scale of the operation has to be vast to make a real contribution: to provide $1\,kW$ per person to a country of 50 million inhabitants needs $3{,}700\,km$ of barrier. And any wave-driven device takes a considerable battering; maintenance is a problem.

Tides. *Tidal* power is power from the moon. Tidal power where tides are high can deliver $3\,W/m^2$ by making both the incoming and the outgoing tidal flow drive turbines. This is about the same as wind power, but few countries are in a position to capture much of it. For those countries with tidal estuaries or at the mouths of large landlocked seas (such as the Mediterranean), harnessing tidal power is an option. But it is not one that will, by itself, provide as much energy as we need.

Geothermal. Geothermal heat is heat conducted up from the hot core of the planet, augmented by heat from the decay of unstable elements in the crust. This heat leaks out at the surface, but not in a useful form. To generate electricity it is necessary to heat water to at least 200°C, and for most of the Earth's crust that means drilling down to about $10\,km$, making it expensive to harvest. In a few places the heat is much closer to the surface, so much so that water bubbling up naturally is above its boiling point. Where this is so (Iceland; hot springs in the United States, New Zealand, and elsewhere), extracting geothermal heat is a practical proposition. For most other countries the contribution it can make is small.

The conclusions: no single renewable source can begin to supply energy on the scale we now use it. A combination of all of them might. But think of the difficulties. There is the low power density, meaning that a large fraction of the area of the country must be dedicated to capturing it. If you cover half the country with solar cells, you cannot also plant crops for biofuel on it, nor can you use it as we now do for agriculture and livestock for food. There is the cost of establishing such a dispersed system and, in the case of offshore wave and wind farms, maintaining it (even on land some 2% of wind turbines are disabled each year by lightning). And there is the opposition, much of it from environmentalists, that paving the country and framing the coast with machinery would create. MacKay's (2008) book examines all this in greater depth; for now we must accept that the dream of copious cheap, pollution-free energy from sun, wind, and wave is not going to become a reality.

10.5 Sustainable materials

No matter how you look at it, using materials costs energy. Let us put that fact aside and examine the degree to which the materials themselves are sustainable. To be so, a material must be drawn from a source that is renewable, either because it grows as fast as we use it or because it reverts to its original state on natural decay and does so in an acceptable time span. For this to be true, the resource and the material must form part of a cycle, like the nitrogen, carbon, or hydrological cycles of the natural world: closed loops that run at steady state, recycling the elements N and C and the compound H_2O such that the resource remains constant.

Which materials meet these constraints? Not many. The obvious examples are wood and natural fibers such as jute, hemp, and cotton. Provided that the total tree stock is constant such that wood is harvested at the same rate as it is grown, wood could be seen as a renewable resource. But today wood is harvested (or simply burned) much faster than it is replaced, making it a diminishing resource, and wood in the form we use it for construction has been cut, dried, chemically treated, and transported, all with some nonrenewable consequences. Similar reservations apply to natural fibers, natural rubber, and leather. Very few of the mater-ials we use today qualify as truly renewable in the sense that those of early man, himself part of a natural ecosystem and its closed-loop cycle, were.

A number of materials do, however, have ingredients that are, for practical purposes, recyclable. The construction industry uses materials in greater quantities than any other, and much construction is in less developed countries where steel, concrete, and fired brick are not easily come by. Architects with concern for the environment and a love of pre-industrial building materials design buildings using an interesting range of near-sustainable materials. Here are some examples:

- *Rammed earth and adobe.* Soil is available almost everywhere. Mix it with straw or hair and a little lime cement, stomp it down between

wooden shuttering, let it set, and remove the shuttering and you
have a wall that will support a light roof. Earth walls have high heat
capacity, useful in a climate that is hot in the day and cold at night.
If the wall is thick enough, its thermal mass keeps the room cool as
the outside temperature rises but warms it at night when the outside
cools. Adobe, in the form of rammed earth bricks, is the traditional
building material of Mexico and parts of Africa. Adobe bricks,
mechanically mixed and extruded, are commercially available today.

■ *Straw and reed.* Straw, a by-product of agriculture, has long been
used in building, but straw bales are a product of modern technology.
Straw bales are like building blocks: stack them up and you create
a wall. Surface them with earth plaster or wood and they become
durable. The walls have low thermal conductivity and low heat
capacity (quite different from rammed earth), so they insulate and
have low thermal mass. Thatch is reed, one with a long history of use
as a roofing material. The reed grows with its base in water so it has
evolved to resist it. Like straw, it insulates and is surprisingly durable;
a well-thatched roof has a life of 80 years.

■ *Hemp and flax.* Hemp and flax are fast-growing grasslike plants
containing fibers of great strength. The fibers have been used since
Roman times for rope and sails, clothing, and construction. The
use of a mix of industrial hemp and lime ("hempcrete") for infill in
wood frame buildings is growing in Europe but is held back elsewhere
(the United States, for instance) because of its mistaken association
with marijuana, derived from a different variety of hemp. The hemp
content of hempcrete, 75% by volume, is truly sustainable, grown
as fast as it is used and requiring no fertilizer. Hempcrete sequesters
0.3 kg of carbon per kg—up to 20 tonnes of carbon in a typical house.

■ *Stone and lime.* Stone may not be renewable, but one might think
the resource from which it is drawn is near infinite. True in general,
but not in particular: carrera marble, Sydney sandstone, Portland
stone, Welsh slate—all are now scarce, even stockpiled, driving up
their price. But more generally, stone is an ecofriendly material,
durable and reliable. "Dressed" stone, the material of city banks,
venerable universities, and corporate head offices, is expensive. Much
labor and quite some energy go into the dressing. Fieldstone is there
to be picked up. "Dry stone" walls are made by skilled stacking of
stones as found. They work well as field boundaries but they are not
load bearing. Stone bonded with a lime mortar, however, is robust
and durable.

■ *Quasi-sustainable materials.* The materials just described are such a limited set that we will have to cast the net wider and consider quasi-renewable materials—those drawn from a resource base so large that, even allowing for exponential growth, there is no risk of exhaustion. Table 10.2 lists the 12 most abundant elements in the Earth's crust, and here we are more fortunate. The list includes many of those that we use in the largest quantities: iron, aluminum, magnesium, and titanium; silicon, sodium, and oxygen, the components of glass and of silicates that form the basis of many ceramics; and carbon and hydrogen, the starting point for all polymers (Figure 2.1). Extracting them economically, of course, requires deposits in which they occur with high enough concentration, but when the total quantity is as enormous as it is for each of these, such deposits are plentiful. It just takes energy.

The "All other" category at the bottom of Table 10.2 is important, too. It includes copper, zinc, tin, lead, cobalt, zirconium, and all the precious

Table 10.2	Abundance of elements in earth's crust
Element	**Abundance in earth's crust by weight (%)**
Oxygen	46.7
Silicon	27.7
Aluminum	8.1
Iron	5.1
Calcium	3.6
Sodium	2.8
Potassium	2.6
Magnesium	2.1
Titanium	0.6
Hydrogen	0.14
Phosphorus	0.13
Carbon	0.09
All others, total	<1

metals and rare earths. Some of these are important in their own right—copper, for instance; others because they are the vitamins, so to speak, of alloy design: small additions to the composition that have a big influence on behavior. For these, rich deposits are leaner and less widely distributed, and as they are depleted we are forced to extract them from ever-leaner ores. If we go lean enough, the total quantities again become large. But that, too, takes energy.

So, the bottom line is *energy*. With enough of it we could continue to use materials as we now do for a long time to come—provided we could afford it and could generate it without poisoning the planet. That is as far as we need to go for now. What we can do about it is the subject of Chapter 11.

10.6 Summary and conclusion

"Sustainable development is development that meets the needs of the present without compromising the ability of future generations to meet their own needs." This much-quoted definition, from the *Brundtland Report* of the World Council on Economic Development (WCED, 1987), captures what most people would agree is the essence of sustainability.

Sustainable development, at the time of writing, is a vision, a grand ideal. Nature achieves it through balanced cycles—the carbon, nitrogen, and water cycles are examples—in which materials are used and, at the end of life, recycled to replenish the source from which they were drawn. That requires a closed loop in which the flows at each point match in rate, for if they do not, waste accumulates somewhere in the cycle. We (the industrial "we") fail to achieve this because the ways in which we use materials either do not form a closed loop, or if they do, the units of the loop do not function at equal rates.

It is clear that we are reaching limits in the way we use fossil fuels for energy and the Earth's resources for materials, both because these become depleted and because the way we use them damages the environment. Sustainable sources of energy exist, but all have the characteristic that the power density is low, so harvesting them requires the dedication of a very large area of land mass or ocean surface. The resources of many of the materials we use in the largest quantities are sufficiently abundant that we can draw on them for a long time to come—provided that we have enough cheap, pollution-free energy to do so. And that, as of now, is what we don't have.

What can we do about it? Forces for change and future options are the subjects of the next chapter, Chapter 11.

10.7 Further reading

Azapagic, A., Perdan, S. and Clift, R. (Eds.) (2004), "Sustainable development in practice", John Wiley. ISBN 0-470-85609-2.

Coulter, S., Bras, B. and Foley, C. (1995), "A lexicon of green engineering terms", Proc. ICED 95, pp. 1–10. *(Coulter presents the nested-box analogy of green design, here used as Figure 10.1.)*

Guidice, F., La Rosa, G. and Risitano, A. (2006), "Product design for the environment", CRC/Taylor and Francis, ISBN 0-8493-2722-9. *(A well-balanced review of current thinking on ecodesign.)*

Lovelock, J. (2000), "Gaia: a new look at life on Earth", Oxford University Press. ISBN 0-19-286218-9. *(A visionary statement of man's place in the environment.)*

MacKay, D.J.C. (2008), "Sustainable energy—without the hot air", UIT Press, Cambridge, UK. ISBN 978-0-9544529-3-3. *(MacKay brings a welcome dose of common sense into the discussion of energy sources and use: fresh air replacing hot air.)*

Nielsen, R. (2005), "The little green handbook", Scribe Publications Pty Ltd, Carlton North. ISBN 1-9207-6930-7. *(A cold-blooded presentation and analysis of hard facts about population, land and water resources, energy, and social trends.)*

Schmidt-Bleek, F. (1997), "How much environment does the human being need? Factor 10: the measure for an ecological economy", Deutscher Taschenbuchverlag. ISBN 3-936279-00-4. *(Both Schmidt-Bleek and von Weizsäcker, referenced below, argue that sustainable development will require a drastic reduction in material consumption.)*

von Weizsäcker, E., Lovins, A.B. and Lovins, L.H. (1997), "Factor four: doubling wealth, halving resource use", Earthscan Publications. ISBN 1-85383-406-8; ISBN-13: 978-1-85383406-6. *(Both von Weizsäcker and Schmidt-Bleek, referenced above, argue that sustainable development will require a drastic reduction in material consumption.)*

WCED (1987), "Report of the World Commission on the Environment and Development", Oxford University Press. *(The so-called Bruntland report launched the current debate and stimulated current actions on moving toward a sustainable existence.)*

Woolley, T. (2006), "Natural building: a guide to materials and techniques", The Crowood Press Limited. ISBN 1-861-26841-6. *(A well-illustrated introduction to traditional building materials and their present-day modifications.)*

10.8 Exercises

E.10.1 Distinguish *pollution control and prevention* (PCP) from *design for the environment* (DFE). When would you use the first? When the second?

E.10.2 What is meant by the *ecological metaphor*? What does it suggest about ways to use materials in a sustainable way?

E.10.3 What are the potential sources of renewable energy? What are the positive and negative aspects of converting to an economy based wholly on renewable sources?

E.10.4 The land area of the Netherlands (Holland) is $41,526 \, km^2$. Its population is 16.5 million, and the average power consumed per capita there is 6.7 kW. If the average wind power is $2 \, W/m^2$ of land area and wind turbines operate at a load factor of 0.5, what fraction of the area of the country would be taken up by turbines to meet the country's energy needs?

E.10.5 The land area of New York state is $131,255 \, km^2$. Its population is 19.5 million, and the average power consumed per capita there is 10.5 kW. If the average wind power is $2 \, W/m^2$ of land area and wind turbines operate at a load factor of 0.25, what fraction of the area of the state would be taken up by turbines to meet its energy needs?

E.10.6 The combined land area of the state of New Mexico ($337,367 \, km^2$) and the state of Nevada ($286,367 \, km^2$) is $623,734 \, km^2$. The population of the United States is 301 million, and the average power consumed per capita there is 10.2 kW. Mass-produced solar cells can capture 10% of the energy that falls on them, which, in New Mexico and Nevada, is roughly $50 W/m^2$. What fraction of the area of the two states would be taken up by solar cells to supply the current needs of the United States?

The bigger picture: future options

11.1 Introduction and synopsis

In drafting the first 11 chapters of this book I have felt, more than once, that I was describing ways to fix a leak in the ceiling while ignoring the flood waters rising through the floor. It's time to look at the bigger picture, and it is not an entirely happy one. Until now we have focused on *facts* and ways to use them, avoiding speculation, judgment, or opinion. Avoiding these when discussing future challenges, particularly those relating to the environment, is more difficult; personal views and informed guesses have a place there. So in this chapter I'm going to say "I" as well as the "we" or "you" I have used so far. The aim of this chapter is to stimulate discussion, not prescribe solutions. You will have your own views. Develop them, but do so in ways that are based on *facts*.

Is this the future? Floods, drought, expanding desserts, and hurricanes. (Images courtesy of Home. vicnet.net.au; Prisonplanet.com; Weathersavvy.com.)

248 CHAPTER 11: The bigger picture: future options

First, a justification for Chapters 1 through 10: there are good reasons for starting in the way we did. It builds on established, accepted methods; it avoids the controversy that plagues much discussion of environmental issues; and it advances understanding. The conclusions reached so far have a solid basis, giving perspective and replacing speculation and misinformation with fact. We have focused on methods to select materials to meet eco-objectives, taking energy and atmospheric carbon as the central actors. Though the gains might be small, it is important to make them; we would fail in our obligations as engineers and scientists *not* to try to do so. Much is learned in this process about where energy goes and where atmospheric carbon comes from. And it helps distinguish material choices that contribute little to atmospheric carbon from those that contribute a great deal; it distinguishes the little fish from the big fish. If we are going to make a real difference, it is the big fish we need to catch.

But is this—will this be—enough? Can it provide a sustainable future? To answer this we must digress a little. If, through change of circumstance, your life is not going well, there are various steps you can take to fix it. The first and normal reaction is to examine the symptoms and try to remedy them with as little disruption to the rest of your life as possible. But if the problem persists, it becomes necessary to look more deeply into the *forces for change* that cause it. If these are real and unstoppable, a greater, more disruptive adjustment will be needed. There is a natural reluctance to do this until you are absolutely sure that the forces are real and the changes unavoidable. It is easier to do nothing, betting that things are not as bad as they seem and that the problem will go away. But if you lose your bet, you are caught unprepared. The adjustments you are then forced to make are not of your choosing. If you like to be in control, *anticipation* is better than *reaction*. This is called the *precautionary principle*.

Using our little story as an analogy for global problems is an oversimplification, but I'm going to make it anyway. The book, thus far, has dealt with minor inconveniences—a little local pollution, a spot of ozone depletion, a little global warming, occasional bits of restrictive legislation—and with corrective measures that disturb the rest of life very little. They can ameliorate the problems, but they will not fix them. Some, at least, of the forces for change are too powerful to be dealt with in that way.

So let us look at the threats, the opportunities, and the options for the future. But first, a question: why do we value materials so little?

11.2 Material value

How is it that materials are seen as of such little value? The waste stream is now so great that densely populated countries are running out of space to store it. We recycle some, but only under duress; it takes legislation, subsidies, and taxes to make us do it. Gold doesn't get thrown away; its intrinsic value protects it. There was a time when the same was true of iron, aluminum, and glass, but not today.

Trends. The three trends plotted in Figure 11.1 give clues. They show, in order, the aggregated price index of a spread of materials,[1] the price of energy,[2] and the purchasing power, expressed as GDP per capita,[3] of developed nations over the last 150 years, all normalized to the dollar value in 2000.[4] In real terms, material prices have decreased steadily over time, making them cheaper than ever before. The price of energy (with some dramatic blips) remained pretty constant until 2008 when, in a very short period, it more than quadrupled, setting back to a value about twice its previous average. Purchasing power, measured by GDP/capita, increased enormously over the same 150 year period. Materials and the goods made from them are cheaper, in real terms, than they have ever been. But though the relative cost of materials has fallen, that of labor has risen, shifting the economic balance away from saving *material* toward that of saving *time*.

If the price of energy rises, what happens to the price of materials? Making them, as we have seen, requires energy; the embodied energy of some is much larger than that of others. The ratio of the cost of this energy to the price of the material is a measure of its sensitivity to energy-price rise. A value near 1 means that energy accounts for nearly all the material price; if energy price doubles, so too will the price of the material. A value of 0.1 means that a doubling of energy price drives the material price up by 10%. Figure 11.2 shows the result. Aluminum and magnesium are energy intensive and thus vulnerable to a change in price. The price of materials such as CFRP has a smaller energy component (here labor and equipment

---

[1] Weighted average of six metals and six nonmetallic commodities. Source: market indicators, *Economist*, Jan. 15, 2000, cited by Lomberg, 2001.

[2] Source: annual energy reviews of the Energy Information Agency (EIA) of the U.S. Department of Energy (www.eia.doe.gov).

[3] GDP per capita in constant 2000$, replotted from data assembled by Lomberg (2001).

[4] Correcting for inflation is not simple. For conversion factors, see www.measuringworth.com.

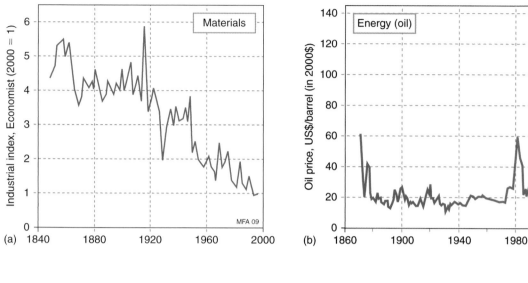

FIGURE 11.1 *The price of (a) materials and (b) oil, and, (c) The GDP per capita, over the last 150 years.*

costs play a larger role), making it less sensitive to energy-price fluctuations. Commodity polymers (PE, PP) lie in between.

Material price and product price. If energy price rises, what happens to the price of products made from particular materials? That depends on the

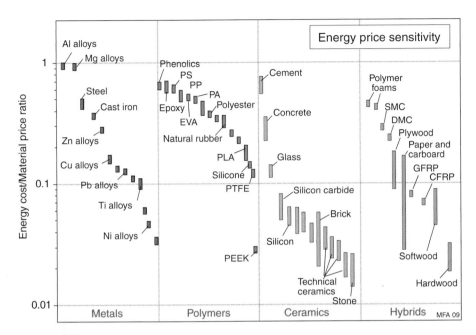

The energy-price sensitivity of materials prices. Those with a value near 1 are the most vulnerable to rising energy prices.

material intensity of the product. Figures 11.3 and 11.4 explain. The vertical axes are the price per unit weight ($/kg) of materials and of products; it gives a common measure by which materials and products can be compared. The measure is a crude one, but has the merit that it is unambiguous and easily determined and it bears some relationship to added value. A product with a price/kg that is only two or three times that of the materials of which it is made is material intensive and is sensitive to material costs; one with a price/kg that is 100 times that of its materials is insensitive to material price. On this scale the price per kg of a contact lens differs from that of a glass bottle by a factor of 10^5, even though both are made of almost the same glass. The cost per kg of a heart valve differs from that of a plastic bottle by a similar factor, even though both are made of polyethylene. There is obviously something to be learned here.

Look first at the price-per-unit weight of materials (see Figure 11.3). The bulk "commodity" materials of construction and manufacture lie in the shaded band; they all cost between $0.05 and $20/kg. Construction materials like concrete, brick, timber, and structural steel lie at the lower end; high-tech materials like titanium alloys lie at the upper. Polymers span a

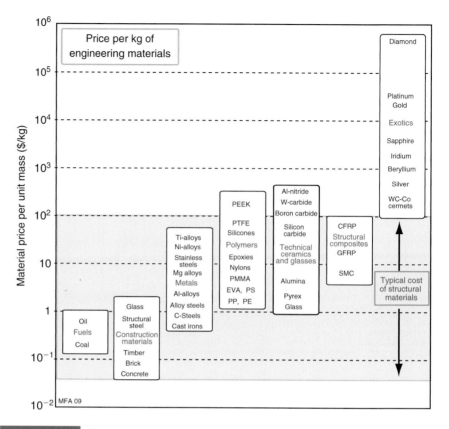

FIGURE 11.3 *The price-per-unit weight diagram for materials. The shaded band spans the range in which lies the most widely used commodity material of manufacture and construction.*

similar range: polyethylene at the bottom, polytetrafluorethylene (PTFE) near the top. Composites lie higher, with GFRP at the bottom and CFRP at the top of the range. Engineering ceramics, at present, lie higher still, though this will change as production increases. Only low-volume "exotic" materials lie much above the shaded band.

The price per kg of products (see Figure 11.4) shows a different distribution. Eight market sectors are shown, covering much of the manufacturing industry. The shaded band on this figure spans the cost of commodity materials, exactly as on the previous figure. Sectors and their products within the shaded band have the characteristic that material cost is a major fraction of product price: up to 50% in civil construction, large marine structures, and some consumer packaging, falling to perhaps 20% as the top of the band is approached (family car—around 25%). The value

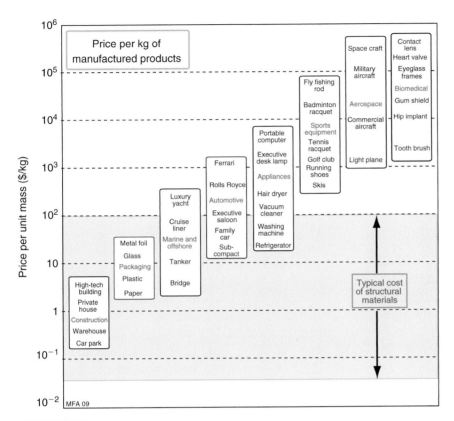

FIGURE 11.4 *The price-per-unit weight diagram for products. The shaded band spans the range in which lies most of the material of which they are made. Products in the shaded band are material intensive; those above it are not.*

added in converting material to product in these sectors is relatively low, but the market volume is large. These constraints condition the choice of materials: they must meet modest performance requirements at the lowest possible cost. The associated market sectors generate a driving force for improved processing of conventional materials to reduce cost without loss of performance or to increase reliability at no increase in cost. For these sectors, incremental improvements in well-tried materials are far more important than revolutionary research findings. Slight improvements in steels, in precision manufacturing methods, or in lubrication technology are quickly assimilated and used.

The products in the upper half of the diagram are technically more sophisticated. The materials of which they are made account for less than

10%—sometimes less than 1%—of the price of the product. The value added to the material during manufacture is high. Product competitiveness is closely linked to material performance. Designers in these sectors have greater freedom in their choice of material and are more willing to adopt them if they have attractive properties. The objective here is performance, with cost as a secondary consideration. These smaller-volume, higher-value-added sectors drive the development of new or improved materials with enhanced performance: materials that are lighter, or stiffer, or stronger, or tougher, or expand less, or conduct better—or all of these at once. They are often energy intensive but are used in such small quantities that this is irrelevant.

The sectors have been ordered to form an ascending sequence, prompting the question: what does the horizontal axis measure? Many factors are involved here, one of which can be identified as "information content." The accumulated knowledge involved in the production of a contact lens or a heart valve is clearly greater than that in a beer glass or a plastic bottle. The sectors on the left make few demands on the materials they employ; those on the right push materials to their limits and at the same time demand the highest reliability. But there are also other factors: market size, competition (or lack of it), perceived value, fashion and taste, and so on. For this reason the diagram should not be over-interpreted; it is a help in structuring information, but it is not a quantitative tool.

11.3 Carbon, energy, and GDP

The greater number of people who use a resource, the faster it is depleted. Global population—symbol P—is rising, and so too is affluence, which we will write as gross domestic product (GDP) per capita, GDP/P. Then material consumption grows as

$$\text{Material consumption} = P \times \frac{GDP}{P} \times \frac{Material}{GDP} \qquad (11.1)$$

The last term is the *material intensity of GDP.* Energy consumption can be expanded in a similar way:

$$\text{Energy consumption} = P \times \frac{GDP}{P} \times \frac{Energy}{GDP} \qquad (11.2)$$

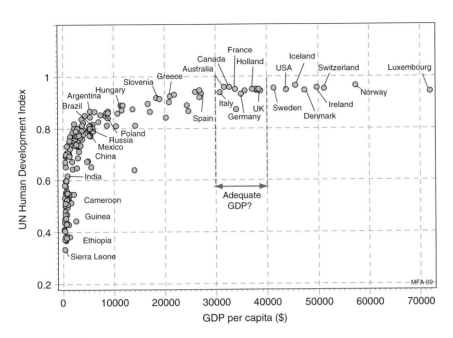

FIGURE 11.5   *The U.N. Human Development Index plotted against GDP/capita.*

Here the last term is the *energy intensity of GDP*. Finally, if we want to focus on carbon emissions rather than energy, we have:

$$\text{Carbon emissions} = P \times \frac{GDP}{P} \times \frac{Energy}{GDP} \times \frac{Carbon}{Energy} \qquad (11.3)$$

This time the last term is the *carbon intensity of energy*.

Any move toward sustainability must stabilize and ultimately reduce all three of the quantities on the left of these equations. That means reducing at least one of the terms on the right. Population forecasts indicate continued growth at least in the near term, and to imagine that the world's population P will not continue to strive for increased affluence GDP/P is, well, unimaginable. You could ask: how much GDP is enough? Figure 11.5 is a plot of one measure of human fulfilment, the U.N. Human Development Index (HDI),[5] as a function of GDP per capita in 2008. There is clearly a plateau. The plot suggests that a GDP per capita between $30,000 and $40,000 is more than enough to achieve it; any further increase buys no

---

[5] The HDI is a compound index that combines provision of education, level of health care, opportunity for individual development, and personal freedom.

increase in HDI. But that, of course, is not how people see it. Attempting to limit GDP per capita is a nonstarter.

So this leaves the last term in each equation as the target. Figure 11.6 is a plot of two of these—the energy/GDP and the CO_2/energy—for 35 developed countries, for electrical energy. The curved contours show the third. If global warming is the concern, it is CO_2/GDP, shown as diagonal contours, that we seek to minimize. The two countries that do best are Sweden and Switzerland; France and Iceland are not far behind. Many other countries have values of some or all of the quantities shown here that are less efficient, some by a factor of 5.

How do the best countries do it? Two, Sweden and Switzerland, have terrain that allows large-scale hydroelectric generation, giving low CO_2, and they have efficient industry, giving low energy/GDP. Iceland has large geothermal capacity, giving low CO_2 again, but it has an energy-intensive economy. Most countries do not have these energy resources. France, without much of either, does well by all three criteria through a combination of nuclear power and efficient manufacture.

There is much more to this, of course. It depends on where the GDP comes from (agriculture, manufacture, trade, finance, tourism, or something else), on how hot or cold the country is (heating and air conditioning), on its natural resources (fertility of soil, mineral, oil, and coal resources),

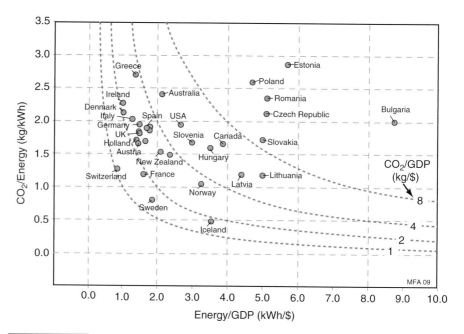

FIGURE 11.6 *The CO_2/energy, energy/GDP, and CO_2/GDP for 33 developed countries.*

and on the way it is governed. But Figure 11.6 provides some perspective: if all the countries plotted here could achieve the performance shown by France, global carbon emissions and energy consumption would fall by a factor of 2 straight away.

11.4 Gathering clouds: threats[6]

Now back to forces for change. Figure 11.7 is the road map for this section and the next. The central spine represents the design or redesign process, moving from market need through the steps of development (including choice of material and process) to the specification and ultimate production of products. The radial boxes summarize, on the left, some of the threats; those on the right, some of the opportunities.

Population. For most of the history of man the population has been small and rising only very slowly (Figure 1.3), but in the last 70 years of the 20th

MFA 09

FIGURE 11.7   *Forces for change: threats on the left, opportunities on the right.*

[6] For full documentation and analysis of the facts listed in this section, see the book by Nielsen (2005) and the IPCC (2007) report listed under Further Reading.

century and the first decade of the 21st the population has exploded, growing from 2 billion to over 6 billion in 80 years. It is expected to rise above 8 billion before stabilizing, and there are already too many to live comfortably on the productive surface area of the planet (Chapter 1).

Energy. Fossil fuels provide almost all the energy we now use (Chapter 2). The consumption has increased by a factor of 14 in the past 100 years and is still rising. Large reserves of easily accessible crude oil and gas are in the hands of a small group of nations; the energy-hungry OECD countries are almost wholly dependent on OPEC for supply. Such a one-sided market carries risk of supply shortages and volatile prices, with threat to economic disruption for the oil-importing nations (see Figure 11.1b).

Water. A second resource, water, is likely to exert an even greater constraint. Of the water on the Earth's surface, only 2.5% is fresh water and much of this is inaccessible, locked up in ice and ground water. The growing population has created water shortages in many parts of the world; one third of the global population lacked adequate supplies at the end of the 20th century. Global warming is expected to make this worse. And without water it is not possible to grow crops or rear livestock.

Land. Industrialized countries need, on average, 6 hectares of productive land per person to support the way they live. The global average is only 1.8 hectares per person, and it is falling further as population continues to grow. The developing nations aspire to a standard of living comparable with those of the developed nations, but at least one resource—land—is insufficient to provide it unless there is a fundamental change in the way we live.

Climate change. We are dumping more than 6 million tons of carbon into the atmosphere each year, pushing the atmospheric concentration from 270 ppm before the Industrial Revolution to above 400 ppm today (Figure 3.8). Global average temperature has risen by 0.75°C since the 19th century. If no further carbon entered the atmosphere (a completely impossible scenario), the inertia of the ecosystem would still result in a further rise of 0.6°C before stabilizing. Increases above 2°C are inevitable. Increasingly precise meteorological modeling gives an idea of the consequences. They include melting of the Arctic ice cap (already evident), rising sea levels, decreasing availability of fresh water, population migration, and loss of species. A total rise of 4°C starts the melting of the Antarctic and Greenland icecaps, with more extreme rise of sea level and flooding of low-lying countries. A total rise of 6°C has predicted consequences that one would rather not think about. All this, of course, is based on modeling, and there are those who dismiss its predictions.[7] The evidence that this progression has already started cannot be ignored. It is worth listening to the modelers; they are an early warning system that did not exist in Malthus's day, giving time to think out the best way to manage these changes.

National security. And finally there is a threat of a different nature. The increased availability and killing power of modern weapons, together with the open nature of Western society, makes it vulnerable to terrorism: the ability of small groups, even individuals, to inflict massive harm. This is a difficult devil to confront when inspired, at least in part, by the large differences in wealth between rich and poor nations and by religious convictions.

11.5 Opportunities

The concerns on the left of Figure 11.7 involve land, climate change, water, and food, but at bottom it is *energy* that is the key to them all. The right side of the figure shows some of the tools we have to deal with them.

Predictive modeling. As we've said, it is better to *anticipate* than to *react*. To foresee a problem is the first step in solving it. Fail to do so and you start with a self-inflicted handicap. No one pretends that meteorological and economic modeling is exact, but both are developed sufficiently to have useful predictive power. The IPCC (2007) *Report on Climate Change* is revealing in this: the many models for the effect of atmospheric carbon on global warming, and the effect of global warming on climate, and the many scenarios that can be run through such models converge on a set of conclusions that are sufficiently robust that it would be foolish not to use them as a basis for anticipation. Modeling is becoming one of the most valuable tools we have for mapping future strategy. One such has to be a transition to carbon-free energy.

Carbon-free energy. As we saw in Chapter 10, it is possible but not easy to generate power on a global scale without burning carbon. Until now the cost of fossil fuel has been so low that there was little incentive to look elsewhere—all the alternatives were too expensive. Shifts to alternative sources, particularly nuclear power, become economically attractive when oil costs over $100 per barrel. But to do so requires new technology and enormous capital investment.

---

[7]See, for example, the provocative but well-researched book by Lomberg (2001).

Science and technology. The scientific and technological resources that could be deployed to develop safe, carbon-free power are enormous. Recall from Chapter 2 that over three quarters of the scientists and engineers who have ever lived are alive and working today; it is one reason that the technologies of information, genetics, surveillance, and defense have developed at the speed they have. They offer a resource that, if deployed to tackle the threats (as I am sure they will, ultimately, be), can do much to ease the pain. It is not clear, though, that technology offers solutions to them all; some problems lie beyond its reach.[8]

The wealth of nations. How will it be paid for? The required investment is huge. But so, too, is the wealth of the developed and the oil-rich nations. Estimates of the cost of the transition to carbon-free energy (see Further Reading for details) vary from 0.5% to 2% of the global GDP, painful but not impossible.

The digital economy. One way to reduce the manmade stress on the environment is to develop ways to use less material and energy per unit of GDP. The digital economy—one in which the trade in information is as central as that in goods—is one way of achieving this goal.

Adaptability. Perhaps the greatest unknown is the extent of human adaptability. Tools and resources exist to plan and implement strategies to deal with the concerns of Figure 11.7. The acceptability of the strategies, however, remains an unknown: can governments and populations be persuaded to adopt them? We should not forget those words of Thomas Malthus, uttered 210 years ago: "the power of population is so superior to the power of the Earth to produce subsistence for man, that premature death must in some shape or other visit the human race." He didn't say when, just that it would. Others, starting from quite different standpoints, have converged on a similar view. The full weight of scientific evidence and of advanced climate modeling now points that way, too. It begins to look as though Malthus might be right. Can we, in the 21st century, find ways to prove him wrong?

11.6 Summary and conclusion

For the past 150 years materials have become cheaper and labor more expensive. The ratio of the two has a profound effect on the way we develop, use, and value materials. In countries where this ratio is still high

---

[8]See, for example, Hardin (1968).

(India, China), they are conserved and recovered. Where it is low (most of the developed world), they are discarded; it is frequently cheaper to buy a new product than to pay the labor cost of having the old one repaired.

Things are now changing. One change is the steep rise in the price of energy, shown in Figure 11.1b; at least part of it is here to stay. And there are others, all acting as drivers for change. The global population has now grown so large that there is insufficient productive land to support it adequately. Global warming caused by atmospheric carbon is causing climate change; allowing it to ramp up further will have harmful effects on health, agriculture, water availability, and weather, all with economic penalties. The dependence on oil, much of it sourced from a few oil-rich countries, has bred a dependence that is increasingly troublesome for the oil-hungry developed nations.

All these will change the ways in which we design with and use materials. All have the effect of making materials more precious. They create incentives to develop materials that better meet the constraints imposed by the design, to care for their health in service, and to cherish them when retired so that they can be retrained, so to speak, to do a new job.

There are many challenges here. They relate both to the materials and to the ways in which we design with them. Exactly how to tackle them is, I think, an interesting topic for further debate.

11.7 Further reading

Hardin, G. (1968), "The tragedy of the commons", Science, Vol. 162, pp. 1243–1248. (*Hardin argues, with convincing examples, that relying on technology alone to solve the problems listed on the left in Figure 11.7—particularly that of population growth—is mistaken. Some problems do not have technical solutions.*)

Kaya, Y. (1990), "Importance of carbon dioxide emission control on GNP growth", Report of the IPCC Energy and Industry Subgroup, Response strategy group. (*The origin of way of breakdown the energy consumption or carbon emissions of GDP into component terms as in equations 11.1 – 11.3*).

Lawson, N. (2008), "An appeal to reason: a cool look at global warming", Duckworth Overlook, London UK. ISBN 978-0-7156-3786-9. (*Nigel Lawson, a distinguished economist and policy maker, argues that almost all currently proposed reactions to the perception of climate change are economically absurd and morally misdirected. The book is interesting for the differences it reveals between the perceptions of scientists and those of political economists – or at least of this political economist.*)

Lomberg, B. (2001), "The skeptical environmentalist: measuring the real state of the world", Cambridge University Press. ISBN 0-521-01068-3. (*A provocative and carefully researched challenge to the now widely held view of the origins and consequences of climate change, helpful in forming your own view of the state of the world.*)

MacKay, D.J.C. (2008), "Sustainable energy—without the hot air", Department of Physics, Cambridge University. www.withouthotair.com. (*MacKay's analysis of the potential for renewable energy is particularly revealing.*)

Malthus, T.R. (1798), "An essay on the principle of population", Printed for Johnson, St. Paul's Church-yard, London. www.ac.wwu.edu/~stephan/malthus/malthus. (*The originator of the proposition that population growth must ultimately be limited by resource availability.*)

Meadows, D.H., Meadows, D.L., Randers, J. and Behrens, W.W. (1997), "The limits to growth", Universe Books. (*The "Club of Rome" report that triggered the first of a sequence of debates in the 20th century on the ultimate limits imposed by resource depletion.*)

Meadows, D.H., Meadows, D.L. and Randers, J. (1992), "Beyond the limits", Earthscan. ISSN 0896-0615. (*The authors of* The Limits to Growth *use updated data and information to restate the case that continued population growth and consumption might outstrip the Earth's natural capacities.*)

Nielsen, R. (2005), "The little green handbook", Scribe Publications Pty Ltd, Carlton North. ISBN 1-9207-6930-7. (*A cold-blooded presentation and analysis of hard facts about population, land and water resources, energy, and social trends.*)

Schmidt-Bleek, F. (1997), "How much environment does the human being need? Factor 10: the measure for an ecological economy", Deutscher Taschenbuchverlag. ISBN 3-936279-00-4. (*Both Schmidt-Bleek and von Weizsäcker, referenced below, argue that sustainable development will require a drastic reduction in material consumption.*)

von Weizsäcker, E., Lovins, A.B. and Lovins, L.H. (1997), "Factor four: doubling wealth, halving resource use", Earthscan Publications. ISBN 1-85383-406-8; ISBN-13: 978-1-85383406-6. (*Both von Weizsäcker and Schmidt-Bleek, referenced above, argue that sustainable development will require a drastic reduction in material consumption.*)

Walker, G. and King, D. (2008), "The hot topic: how to tackle global warming and still keep the lights on", Bloomsbury Publishing. ISBN 9780-7475-9395-9. (*A readable paraphrase of the IPCC (2007)* Report on Climate *Change, with discussion of the political obstacles to finding solutions—a topic on which Dr. King is an expert; he was, for some years, chief science advisor to the U.K. Government.*)

11.8 Exercises

E.11.1. An aluminum saucepan weighs 1.2 kg and costs $10. The embodied energy of aluminum is 220 MJ/kg. If the cost of industrial electric power doubles from 0.0125 $/MJ to 0.025 $/MJ, how much will it change the cost of the saucepan?

E.11.2. An MP3 player weighs 100 grams and costs $120. The embodied energy of assembled integrated electronics of this sort is about 2000 MJ/kg. If the cost of industrial electric power doubles from 0.0125 $/MJ to 0.025 $/MJ, how much will it change the cost of the MP3 player?

E.11.3 Cars in Cuba are repaired and continue to be used when 25 years old. The average life of a car in the United States is 13 years. What is the underlying reason for this?

E.11.4 Use the Worldwide Web to research the meaning and history of "The precautionary principle". Select and report the definition that, in your view, best sums up the meaning.

E.11.5 Use the Worldwide Web to research examples of problems that are approached by predicitive modeling.

Material profiles

CONTENTS

12.1 Introduction and synopsis

You can't calculate anything without numbers. This chapter provides them. It takes the form of double-page data sheets for 47 of the materials used in the greatest quantities in modern products. The sheets list the annual production and reserves, the embodied energy, process energies, and the carbon footprints associated with these. They list, too, the general, mechanical, thermal, and electrical properties, important because it is these that determine the environmental consequences of the use phase of life. And they provide basic information about recycling at end of life.

Each section starts with a brief introduction to a material family: *metals and alloys, polymers and elastomers, ceramics and glasses*, and *hybrids* (composites, foams, and natural materials). Within a section the material

Table 12.1	The Material Profiles		
Metals and alloys	Polymers and elastomers	Ceramics and glasses	Hybrids: composites, foams, and natural materials
Aluminum alloys	ABS	Brick	CFRP
Magnesium alloys	Polyamide PA	Stone	GFRP
Titanium alloys	Polypropylene, PP	Concrete	Sheet molding compound
Copper alloys	Polyethylene, PE	Alumina	Bulk molding compound
Lead alloys	Polycarbonate, PC	Soda-lime glass	Rigid polymer foam
Zinc alloys	PET	Borosilicate glass	Flexible polymer foam
Nickel-chrome alloys	PVC		Paper and cardboard
Nickel-based superalloys	Polystyrene, PS		Plywood
Low carbon steel	Polylactide, PLA		Softwood, along grain
Low alloy steel	PHB		Softwood, across grain
Stainless steel	Epoxy		Hardwood, along grain
Cast iron	Polyester		Hardwood, across grain
	Phenolic		
	Natural rubber, NR		
	Butyl rubber, BR		
	EVA		
	Polychloroprene, CR		

profiles appear in the order shown in Table 12.1. Each data sheet has a description and an image of the material in use. The data that follows are listed as *ranges* spanning the typical spread of values of the property. When a single ("point") value is needed for exercises or projects, it is best to use the geometric mean of the two values listed on the sheet.[1]

A warning. The engineering properties of materials—their mechanical, thermal and electrical attributes—are well characterized. They are measured with sophisticated equipment according to internationally accepted Standards and are reported in widely accessible handbooks and databases. They are not *exact*, but their precision—when it matters—is reported; many are known to 3-figure accuracy, some to more.

The eco-properties of materials are not like that. There are no sophisticated test-machines to measure embodied energies or carbon footprints. International standards, detailed in ISO 14040 and discussed in Chapter 3, lay out procedures, but these are vague and not easily applied. The differences in the process routes by which materials are made in different production facilities, the difficulty in setting system boundaries and the procedural problems in assessing energy, CO_2 and the other eco-attributes all contribute to the imprecision.

---

[1] The geometric mean of two numbers X and Y is $\sqrt{XY}$.

So just how far can values for eco-properties be trusted? An analysis, documented in Section 6.3, suggests a standard deviation of $\pm 10\%$ at best. To be significantly different, values of eco-properties must differ by *at least* 20%. The difference between materials with really large and really small values of embodied energy or carbon footprint is a factor of 1000 or more, so the imprecision still allows firm distinctions to be drawn. But when the differences are small, other factors such as the recycle content of the material, its durability (and thus life-time) and the ability to recycle it at end of life are far more significant in making the selection.

12.2 Metals and alloys

Most of the elements in the periodic table are metals. Metals have "free" electrons—electrons that flow in an electric field—so they conduct electricity well, they reflect light, and, viewed with the light behind them, they are opaque. The metals used in product design are, almost without exception, alloys. Steels (iron with carbon and a host of other alloying elements to make them harder, tougher, or more corrosion resistant) account for more than 90% of all the metals consumed in the world; aluminum comes next, followed by copper, nickel, zinc, titanium, magnesium, and tungsten, in the order that was illustrated in Figure 2.1.

Compared to all other classes of material, metals are stiff, strong, and tough, but they are heavy. They have relatively high melting points, allowing some metal alloys to be used at temperatures as high as 2200°C. Only one metal—gold—is chemically stable as a metal; all the others will, given the chance, react with oxygen, sulfur, phosphorous, or carbon to form compounds that are more stable than the metal itself, making them vulnerable to corrosion. There are numerous ways of preventing or slowing this corrosion to an acceptable level, but they require maintenance. Metals are ductile, allowing them to be shaped by rolling, forging, drawing, and extrusion; they are easy to machine with precision; and they can be joined in many different ways. This allows a flexibility of design with metals that is only now being challenged by polymers.

Primary production of metals is energy intensive. Many, among them aluminum, magnesium, and titanium, require at least twice as much energy per unit weight (or five times more per unit volume) as commodity polymers. But most metals can be recycled efficiently, and the energy required to do so is much less than that required for primary production. Some are toxic, particularly the heavy metals—lead, cadmium, mercury. Some, however, are so inert that they can be implanted in the body: stainless steels, cobalt alloys, and certain alloys of titanium are examples. Here are data sheets for the 12 metals and alloys in the order in which they are listed Table 12.1.

Aluminum alloys

The material. Aluminum was once so rare and precious that the Emperor Napoleon III of France had a set of cutlery made from it that cost him more than silver. But that was 1860; today, nearly 150 years later, aluminum spoons are things you throw away—a testament to our ability to be both technically creative and wasteful. Aluminum, the first of the "light alloys" (with magnesium and titanium), is the third most abundant metal in the Earth's crust (after iron and silicon), but extracting it costs much energy. It has grown to be the second most important metal in the economy (steel comes first) and the mainstay of the aerospace industry.

Composition
Al + alloying elements, e.g., Mg, Mn, Cr, Cu, Zn, Zr, Li

General properties
Density	2500 – 2900	kg/m^3
Price	2.5 – 2.8	USD/kg

Mechanical properties
Young's modulus	68 – 82	GPa
Yield strength (elastic limit)	30 – 550	MPa
Tensile strength	58 – 550	MPa
Elongation	1 – 44%	
Hardness—Vickers	12 – 150	HV
Fatigue strength at 10^7 cycles	22 – 160	MPa
Fracture toughness	22 – 35	MPa.m$^{1/2}$

Thermal properties
Melting point	495 – 640	°C
Service temperature	120 – 200	°C Maximum
Thermal conductor or insulator?	Good conductor	
Thermal conductivity	76 – 240	W/m.K
Specific heat capacity	860 – 990	J/kg.K
Thermal expansion coefficient	21 – 24	µstrain/°C

Electrical properties
Electrical conductor or insulator?	Good conductor	
Electrical resistivity	2.5 – 6	µohm.cm

Cast and wrought aluminum alloys, examples of the wide range of properties of this, the most widely used light alloy.

Ecoproperties: material

Annual world production	33×10^6	– 34×10^6	tonne/yr
Reserves	20×10^9	– 2.2×10^9	tonne
Embodied energy, primary production	200	– 240	MJ/kg
CO_2 footprint, primary production	11	– 13	kg/kg
Water usage	*125	– 375	l/kg
Eco-indicator	740	– 820	millipoints/kg

Ecoproperties: processing

Casting energy	*2.4	– 2.9	MJ/kg
Casting CO_2 footprint	*0.14	– 0.17	kg/kg
Deformation processing energy	*2.4	– 2.9	MJ/kg
Deformation processing CO_2 footprint	*0.19	– 0.23	kg/kg

Recycling

Embodied energy, recycling	18	– 21	MJ/kg
CO_2 footprint, recycling	1.1	– 1.2	kg/kg
Recycle fraction in current supply	33	– 55	%

Typical uses. Aerospace engineering; automotive engineering—pistons, clutch housings, exhaust manifolds; sports equipment such as golf clubs and bicycles; die-cast chassis for household and electronic products; siding for buildings; reflecting coatings for mirrors; foil for containers and packaging; beverage cans; electrical and thermal conductors.

Magnesium alloys

The material. Magnesium is a metal almost indistinguishable from aluminum in color but of lower density. It is the lightest of the light-metal trio (with partners aluminum and titanium), and light it is: a computer case made from magnesium is barely two thirds as heavy as one made from aluminum. Titanium, aluminum, and magnesium are the mainstays of airframe engineering. Only beryllium is lighter, but its expense and potential toxicity limit its use to special applications only. Magnesium is flammable, but this is only a problem when it is in the form of powder or very thin sheet. It costs more than aluminum but nothing like as much as titanium.

Composition
Mg + alloying elements, e.g., Al, Mn, Si, Zn, Cu, Li, rare earth elements.

General properties

Density	1740 – 1950	kg/m^3
Price	*4.5 – 5.0	USD/kg

Mechanical properties

Young's modulus	42 – 47	GPa
Yield strength (elastic limit)	70 – 400	MPa
Tensile strength	185 – 475	MPa
Elongation	3.5 – 18	%
Hardness—Vickers	35 – 135	HV
Fatigue strength at 10^7 cycles	*60 – 225	MPa
Fracture toughness	*12 – 18	MPa.m$^{1/2}$

Thermal properties

Melting point	447 – 649	°C
Maximum service temperature	120 – 200	°C
Thermal conductor or insulator?	Good conductor	
Thermal conductivity	50 – 156	W/m.K
Specific heat capacity	955 – 1060	J/kg.K
Thermal expansion coefficient	24.6 – 28	µstrain/°C

Electrical properties

Electrical conductor or insulator?	Good conductor	
Electrical resistivity	4.15 – 15	µohm.cm

Magnesium, the lightest of the light alloys, is increasingly used for components of cars and other vehicles.

Ecoproperties: material

Annual world production	645×10^3	– 655×10^3	tonne/yr
Reserves	1×10^9	– 1.1×10^9	tonne
Embodied energy, primary production	356	– 394	MJ/kg
CO_2 footprint, primary production	22.4	– 24.8	kg/kg
Water usage	*500	– 1500	l/kg

Ecoproperties: processing

Casting energy	*2.1	– 2.6	MJ/kg
Casting CO_2 footprint	*0.13	– 0.16	kg/kg
Deformation processing energy	*3.8	– 4.6	MJ/kg
Deformation processing CO_2 footprint	*0.3	– 0.37	kg/kg

Recycling

Embodied energy, recycling	8.4	– 9.3	MJ/kg
CO_2 footprint, recycling	1.8	– 1.95	kg/kg
Recycle fraction in current supply	9.2	– 11.4	%

Typical uses. Aerospace; automotive; sports goods such as bicycles; nuclear fuel cans; vibration damping and shielding of machine tools; engine case castings; crank cases; transmission housings; automotive wheels; ladders; housings for electronic equipment, particularly mobile phone and portable computer chassis; camera bodies; office equipment; marine hardware and lawnmowers.

Titanium alloys

The material. The alloys of titanium have the highest strength-to-weight ratio of any structural metal, about 25% greater than the best alloys of aluminum or steel. Titanium alloys can be used at temperatures up to 500°C; compressor blades of aircraft turbines are made of them. They have unusually poor thermal and electrical conductivity and low expansion coefficients. The alloy Ti 6% Al 4% V is used in quantities that exceed those of all other titanium alloys combined. The data in this record describes it and similar alloys.

Composition
Ti + alloying elements, e.g., Al, Zr, Cr, Mo, Si, Sn, Ni, Fe, V.

General properties
Density	4400 – 4800	kg/m^3
Price	96.9 – 107	USD/kg

Mechanical properties
Young's modulus	110 – 120	GPa
Yield strength (elastic limit)	750 – 1200	MPa
Tensile strength	800 – 1450	MPa
Elongation	5 – 10	%
Hardness—Vickers	267 – 380	HV
Fatigue strength at 10^7 cycles	*589 – 617	MPa
Fracture toughness	55 – 70	MPa.m$^{1/2}$

Thermal properties
Melting point	1480 – 1680	°C
Maximum service temperature	450 – 500	°C
Thermal conductor or insulator?	Poor conductor	
Thermal conductivity	7 – 14	W/m.K
Specific heat capacity	645 – 655	J/kg.K
Thermal expansion coefficient	8.9 – 9.6	μstrain/°C

Electrical properties
Electrical conductor or insulator?	Good conductor	
Electrical resistivity	100 – 170	μohm.cm

Adiabatic heating heats the air in the compressor to about 500°C, requiring the use of titanium alloys for the blades. (Reproduced with the permission of Rolls-Royce plc, © Rolls-Royce plc 2004.)

Ecoproperties: material

Annual world production	120×10^5	– 122×10^3	tonne/yr
Reserves	50×10^6	– 52×10^6	tonne
Embodied energy, primary production	600	– 740	MJ/kg
CO_2 footprint, primary production	*38	– 44	kg/kg
Water usage	*470	– 1410	l/kg

41   ≈ 57.25

Ecoproperties: processing

Casting energy	5.02	– 5.77	MJ/kg
Casting CO_2	0.301	– 0.346	kg/kg
Forging, rolling energy	*4.71	– 5.7	MJ/kg
Forging, rolling CO_2	*0.377	– 0.456	kg/kg

.325
.415
≈ .75

Recycling

Embodied energy, recycling	228	– 281	MJ/kg
CO_2 footprint, recycling	*14.4	– 16.7	kg/kg
Recycle fraction in current supply	20	– 24	%

≈ 15.5

Typical uses. Aircraft turbine blades; general aerospace applications; chemical engineering; pressure vessels; high-performance automotive parts such as connecting rods; heat exchangers; bioengineering; medical; missile fuel tanks; compressors; valve bodies; light springs, surgical implants; marine hardware, paper-pulp equipment; sports equipment such as golf clubs and bicycles.

680.5

254.5

935

Copper alloys

The material. In Victorian times people washed their clothes in a "copper"—a vat or tank of beaten copper sheet, heated over a fire. The device exploited both the high ductility and the thermal conductivity of the material. Copper has a distinguished place in the history of civilization: it enabled the technology of the Bronze Age (3000–1000 BC). It is used in many forms: as pure copper, as copper-zinc alloys (brasses), as copper-tin alloys (bronzes), and as copper-nickel and copper-beryllium. The designation of *copper* is used when the percentage of copper is more than 99.3%.

Composition
Cu with up to 40% Zn or 30% Sn, Al or Ni.

General properties

Density	8930 – 9140	kg/m^3
Price	6.7 – 7.4	USD/kg

Mechanical properties

Young's modulus	112 – 148	GPa
Yield strength (elastic limit)	30 – 350	MPa
Tensile strength	100 – 400	MPa
Elongation	3 – 50	%
Hardness—Vickers	44 – 180	HV
Fatigue strength at 10^7 cycles	60 – 130	MPa
Fracture toughness	30 – 90	MPa.m$^{1/2}$

Thermal properties

Melting point	982 – 1080	°C
Maximum service temperature	180 – 300	°C
Thermal conductor or insulator?	Good conductor	
Thermal conductivity	160 – 390	W/m.K
Specific heat capacity	372 – 388	J/kg.K
Thermal expansion coefficient	16.9 – 18	μstrain/°C

Electrical properties

Electrical conductor or insulator?	Good conductor	
Electrical resistivity	1.74 – 5.01	μohm.cm

Copper and brass are exceptionally ductile and can be worked into complex shapes.

Ecoproperties: material

Annual world production	15×10^6	– 16×10^6	tonne/yr
Reserves	470×10^6	– 490×10^6	tonne
Embodied energy, primary production	68	– 74	MJ/kg
CO_2 footprint, primary production	4.9	– 5.6	kg/kg
Water usage	*150	– 450	l/kg
Eco-indicator	1300	– 1500	millipoints/kg

Ecoproperties: processing

Casting energy	*2.5	– 2.8	MJ/kg
Casting CO_2 footprint	*0.15	– 0.17	kg/kg
Deformation processing energy	*1.8	– 2.2	MJ/kg
Deformation processing CO_2 footprint	*0.14	– 0.16	kg/kg

Recycling

Embodied energy, recycling	17	– 18.5	MJ/kg
CO_2 footprint, recycling	1.2	– 1.4	kg/kg
Recycle fraction in current supply	40	– 60	%

Typical uses. Electrical wiring, cables, bus bars, high strength, high conductivity wires and sections, overheads lines, contact wires, resistance-welding electrodes, terminals, high-conductivity items for use at raised temperatures, heat exchangers, coinage, pans, kettles and boilers, plates for etching and engraving, roofing and architecture, cast sculptures, pumps, valves, marine propellers.

Lead alloys

The material. When the Romans conquered Britain in 43 AD they discovered rich deposits of lead ore and started a mining and refining industry that was to continue for 1000 years (the symbol for lead, Pb, derives from its Latin name: *plumbum*). They used it for pipes, cisterns, and roofs, this last a use that continues to the present day. The biggest single use of lead (70% of the total) is as electrodes in lead acid batteries.

Composition
Pb + 0 to 25% Sb or 0 to 60 % Sn, sometimes with some Ca.

General properties

Density	11300 – 11400	kg/m^3
Price	*6.0 – 6.6	USD/kg

Mechanical properties

Young's modulus	12.5 – 15	GPa
Yield strength (elastic limit)	8 – 14	MPa
Tensile strength	12 – 20	MPa
Elongation	30 – 60	%
Hardness—Vickers	3 – 6.5	HV
Fatigue strength at 10^7 cycles	2 – 9	MPa
Fracture toughness	*5 – 15	MPa.m$^{1/2}$

Thermal properties

Melting point	183 – 31	°C
Maximum service temperature	*70 – 120	°C
Thermal conductor or insulator?	Good conductor	
Thermal conductivity	22 – 36	W/m.K
Specific heat capacity	122 – 145	J/kg.K
Thermal expansion coefficient	18 – 32	μstrain/°C

Lead weathers well and is exceptionally durable and corrosion resistant.

Electrical properties

Electrical conductor or insulator?	Good conductor		
Electrical resistivity	20	– 22	μohm.cm

Ecoproperties: material

Annual world production	3.3×10^6	– 3.4×10^6	tonne/yr
Reserves	66×10^6	– 67×10^6	tonne
Embodied energy, primary production	53	– 58	MJ/kg
CO_2 footprint, primary production	3.3	– 3.7	kg/kg
Water usage	*175	– 525	l/kg
Eco-indicator	610	– 670	millipoints/kg

Handwritten annotations: 67.5 ; 56 ; 3.5 ; ≈4.1

Ecoproperties: processing

Casting energy	*0.4	– 0.49	MJ/kg
Casting CO_2 footprint	*0.024	– 0.03	kg/kg
Deformation processing energy	*1.6	– 2.7	MJ/kg
Deformation processing CO_2 footprint	*0.13	– 0.14	kg/kg

Handwritten annotations: .45 ; .027 ; 2.25 ; .135 ; ≈ .16

Recycling

Embodied energy, recycling	8.4	– 9.2	MJ/kg
CO_2 footprint, recycling	0.46	– 0.51	kg/kg
Recycle fraction in current supply	70	– 75	%

Handwritten annotations: 8.8 ; .49.5

Typical uses. Roofs, wall cladding, pipe work, window seals, and flooring in buildings; sculpture and table wear as pewter; solder for electrical circuits and for mechanical joining, bearings, printing type, ammunition, pigments, X-ray shielding, corrosion-resistant material in the chemical industry and electrodes for lead acid batteries.

Zinc die-casting alloys

The material. Zinc is a bluish-white metal with a low melting point (420°C). The slang in French for a bar or pub is *le zinc*; bar counters in France used to be clad in zinc—many still are—to protect them from the ravages of wine and beer. Bar surfaces have complex shapes: a flat top, curved profiles, rounded or profiled edges. These two sentences say much about zinc; it is ductile; it is hygienic; it survives exposure to acids (wine), to alkalis (cleaning fluids), and to misuse (upset customers). These remain among the reasons it is still used today. Another is the "castability" of zinc alloys; their low melting point and fluidity give them a leading place in die casting.

Composition
Zn + 3–30% Al, typically, often with up to 3% Cu.

General properties

Density	4950	– 7000	kg/m^3
Price	*3.09	– 3.4	USD/kg

Mechanical properties

Young's modulus	68	– 100	GPa
Yield strength (elastic limit)	80	– 450	MPa
Tensile strength	135	– 510	MPa
Compressive strength	80	– 450	MPa
Elongation	1	– 30	%
Hardness—Vickers	55	– 160	HV
Fatigue strength at 10^7 cycles	*20	– 160	MPa
Fracture toughness	*10	– 70	MPa.m$^{1/2}$

Thermal properties

Melting point	375	– 492	°C
Maximum service temperature	*80	– 110	°C
Thermal conductor or insulator?	Good conductor		
Thermal conductivity	100	– 130	W/m.K
Specific heat capacity	405	– 535	J/kg.K
Thermal expansion coefficient	23	– 28	µstrain/°C

Electrical properties

Electrical conductor or insulator?	Good conductor		
Electrical resistivity	5.4	– 7.2	µohm.cm

Zinc die castings are cheap, have high surface finish, and can be complex in shape. On the left, a corkscrew; everything except the screw itself is die-cast zinc alloy. On the right, a carburetor body.

Ecoproperties: material

Annual world production	9.7×10^6	– 1.0×10^7	tonne/yr
Reserves	2.18×10^8	– 2.21×10^8	tonne
Embodied energy, primary production	70	– 75	MJ/kg
CO_2 footprint, primary production	3.7	– 4	kg/kg
Water usage	*160	– 521	l/kg
Eco-indicator	3000	– 3400	millipoints/kg

Ecoproperties: processing

Casting energy	*1.09	– 1.32	MJ/kg
Casting CO_2	*0.065	– 0.08	kg/kg

Recycling

Embodied energy, recycling	12.6	– 13.5	MJ/kg
CO_2 footprint, recycling	0.66	– 0.72	kg/kg
Recycle fraction in current supply	20	– 25	%

Typical uses. Die castings; automotive parts and tools; gears; household goods; office equipment; building hardware; padlocks; toys; business machines; sound reproduction equipment; hydraulic valves; pneumatic valves; soldering; handles; gears; automotive components.

Nickel-chromium alloys

The material. Nickel forms a wide range of alloys, valued by the chemical engineering and food-processing industries for their resistance to corrosion and by the makers of furnaces and high temperature equipment for their ability to retain useful strength at temperatures up to 1200°C. Typical of these are the nickel-chromium (Ni-Cr) alloys, often containing some iron (Fe) as well. The chromium increases the already good resistance to corrosion and oxidation by creating a surface film of Cr_2O_3, the same film that makes stainless steel stainless. The data given here is for nickel-chromium alloys. There are separate records for stainless steel and nickel-based superalloys.

Composition
Ni + 10 to 30% Cr + 0 to 10% Fe.

General properties

Density	8300	– 8500	kg/m^3
Price	*34.5	– 37.9	USD/kg

Mechanical properties

Young's modulus	200	– 220	GPa
Yield strength (elastic limit)	365	– 460	MPa
Tensile strength	615	– 760	MPa
Elongation	20	– 35	%
Hardness—Vickers	160	– 200	HV
Fatigue strength at 10^7 cycles	*245	– 380	MPa
Fracture toughness	*80	– 110	MPa.m$^{1/2}$

Thermal properties

Melting point	1350	– 1430	°C
Maximum service temperature	*900	– 1000	°C
Thermal conductor or insulator?	Poor conductor		
Thermal conductivity	9	– 15	W/m.K
Specific heat capacity	430	– 450	J/kg.K
Thermal expansion coefficient	12	– 14	µstrain/°C

Electrical properties

Electrical conductor or insulator?	Good conductor		
Electrical resistivity	102	– 114	µohm.cm

The heating elements of the dryer and toaster are Nichrome, an alloy of nickel and chromium that resists oxidation well.

Ecoproperties: material

Annual world production	1.5×10^6	–	1.6×10^6	tonne/yr
Reserves	63×10^6	–	65×10^6	tonne
Embodied energy, primary production	127	–	140	MJ/kg
CO_2 footprint, primary production	*7.89	–	8.82	
Water usage	*134	–	512	l/kg
Eco-indicator	4900	–	5500	millipoints/kg

Ecoproperties: processing

Casting energy	*3.68	–	4.06	MJ/kg
Casting CO_2	*0.221	–	0.243	kg/kg
Forging, rolling energy	*2.52	–	2.78	MJ/kg
Forging, rolling CO_2	*0.202	–	0.222	kg/kg

Recycling

Embodied energy, recycling	31.8	–	35	MJ/kg
CO_2 footprint, recycling	*1.97	–	2.21	kg/kg
Recycle fraction in current supply	22	–	26	%

Typical uses. Heating elements and furnace windings; bimetallic strips; thermocouples; springs; food-processing equipment; chemical engineering equipment.

Nickel-based superalloys

The material. With a name like *superalloy* there has to be something special here. There is. Superalloy is a name applied to nickel-based, iron-based, and cobalt-based alloys that combine exceptional high-temperature strength with excellent corrosion and oxidation resistance. Without them, jet engines would not be practical: they can carry load continuously at temperatures up to 1200°C. The nickel-based superalloys are the ultimate metallic cocktail: nickel with a good slug of chromium and lesser shots of cobalt, aluminum, titanium, molybdenum, zirconium, and iron. The data in this record spans the range of high-performance, nickel-based superalloys.

Composition
Ni + 10 to 25% Cr + Ti, Al, Co, Mo, Zr, B, and Fe in varying proportions.

General properties
Density	7750 – 8650	kg/m^3
Price	*28.5 – 31.4	USD/kg

Mechanical properties
Young's modulus	150 – 245	GPa
Yield strength (elastic limit)	300 – 1.9e3	MPa
Tensile strength	400 – 2.1e3	MPa
Compressive strength	300 – 1.9e3	MPa
Elongation	0.5 – 60	%
Hardness—Vickers	200 – 600	HV
Fatigue strength at 10^7 cycles	*135 – 900	MPa
Fracture toughness	65 – 110	MPa.m$^{1/2}$

Thermal properties
Melting point	1280 – 1410	°C
Maximum service temperature	*900 – 1200	°C
Thermal conductor or insulator?	Good conductor	
Thermal conductivity	8 – 17	W/m.K
Specific heat capacity	380 – 490	J/kg.K
Thermal expansion coefficient	9 – 16	μstrain/°C

Electrical properties
Electrical conductor or insulator?	Poor conductor	
Electrical resistivity	84 – 240	μohm.cm

Nickel is the principal ingredient of superalloys used for high-temperature turbines and chemical engineering equipment. On the left, a gas turbine (image courtesy of Kawasaki Turbines). On the right, a single superalloy blade.

Ecoproperties: material

Annual world production	1.5×10^6	– 1.6×10^6	tonne/yr
Reserves	6.3×10^7	– 6.5×10^7	tonne
Embodied energy, primary production	135	– 150	MJ/kg
CO_2 footprint, primary production	*7.89	– 9.2	kg/kg
Water usage	*134	– 484	l/kg
Eco-indicator	4900	– 5500	millipoints/kg

Ecoproperties: processing

Casting energy	*3.38	– 4.09	MJ/kg
Casting CO_2	*0.20	– 0.24	kg/kg
Forging, rolling energy	*3.26	– 3.95	MJ/kg
Forging, rolling CO_2	*0.26	– 0.31	kg/kg
Metal powder forming energy	*11.3	– 13.6	MJ/kg
Metal powder forming CO_2	*0.90	– 1.09	kg/kg

Recycling

Embodied energy, recycling	33.8	– 37.5	MJ/kg
CO_2 footprint, recycling	*1.97	– 2.3	kg/kg
Recycle fraction in current supply	22	– 26	%

Typical uses. Blades, disks, and combustion chambers in turbines and jet engines; rocket engines; general structural aerospace applications; light springs, high-temperature chemical engineering equipment; bioengineering and medical.

Low carbon steel

The material. Think of steel and you think of railroads, oil rigs, tankers, and skyscrapers. And what you are thinking of is not just steel, it is carbon steel. That is the metal that made them possible; nothing else is at the same time so strong, so tough, so easily formed, and so cheap. Carbon steels are alloys of iron with carbon and often a little manganese, nickel, and silicon. Low carbon or "mild" steels have the least carbon—less than 0.25%. They are relatively soft, easily rolled to plate, I-sections or rod (for reinforcing concrete), and are the cheapest of all structural metals; it is these that are used on a huge scale for reinforcement, steel-framed buildings, ship plate, and the like.

Composition
Fe/0.02–0.3C.

General properties

Density	7800	– 7900	kg/m^3
Price	*0.79	– 0.90	USD/kg

Mechanical properties

Young's modulus	200	– 215	GPa
Yield strength (elastic limit)	250	– 395	MPa
Tensile strength	345	– 580	MPa
Elongation	26	– 47	%
Hardness—Vickers	107	– 172	HV
Fatigue strength at 10^7 cycles	*203	– 293	MPa
Fracture toughness	*41	– 82	MPa.m$^{1/2}$

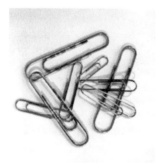

Thermal properties

Melting point	1480	– 1530	°C
Maximum service temperature	*350	– 400	°C
Thermal conductor or insulator?	Good conductor		
Thermal conductivity	49	– 54	W/m.K
Specific heat capacity	460	– 505	J/kg.K
Thermal expansion coefficient	11.5	– 13	μstrain/°C

Electrical properties

Electrical conductor or insulator?	Good conductor		
Electrical resistivity	15	– 20	μohm.cm

Ecoproperties: material

Annual world production	1.1×10^9	– 1.2×10^9	tonne/yr
Reserves	78×10^9	– 79×10^9	tonne
Embodied energy, primary production	29	– 35	MJ/kg
CO_2 footprint, primary production	2.2	– 2.8	kg/kg
Water usage	*23	– 69	l/kg
Eco-indicator	79	– 87	millipoints/kg

Ecoproperties: processing

Casting energy	*4.2	– 4.6	MJ/kg
Casting CO_2 footprint	*0.25	– 0.28	kg/kg
Deformation processing energy	*2.2	– 2.6	MJ/kg
Deformation processing CO_2 footprint	*0.18	– 0.21	kg/kg

Recycling

Embodied energy, recycling	8.1	– 9.8	MJ/kg
CO_2 footprint, recycling	0.6	– 0.8	kg/kg
Recycle fraction in current supply	40	– 44	%

Typical uses. Low carbon steels are used so widely that no list would be complete. Reinforcement of concrete, steel sections for construction, sheet for roofing, car-body panels, cans, and pressed-sheet products give an idea of the scope.

Low alloy steel

The material. Addition of manganese (Mn), nickel (Ni), molybdenum (Mo), or chromium (Cr) to steel lowers the critical quench rate and comes to create martensite, allowing thick sections to be hardened and then tempered. Adding some vanadium, V, as well creates a dispersion of carbides, giving strength while retaining toughness and ductility. Chrome-molybdenum steels such as AIS 4140 are used for aircraft tubing and other high-strength parts. Chrome-vanadium steels are used for crank and propeller shafts and high-quality tools. Steels alloyed for this purpose are called *low alloy steels*, and the property they have is called *hardenability*.

Composition
Fe/ < 1.0 C/ < 2.5 Cr/ < 2.5 Ni/ < 2.5 Mo/ < 2.5 V.

General properties
Density	7800	– 7900	kg/m^3
Price	*1.0	– 1.1	USD/kg

Mechanical properties
Young's modulus	205	– 217	GPa
Yield strength (elastic limit)	400	– 1500	MPa
Tensile strength	550	– 1760	MPa
Elongation	3	– 38	%
Hardness—Vickers	140	– 692	HV
Fatigue strength at 10^7 cycles	*248	– 700	MPa
Fracture toughness	14	– 200	MPa.m$^{1/2}$

Thermal properties
Melting point	1380	– 1530	°C
Maximum service temperature	*500	– 550	°C
Thermal conductor or insulator?	Good conductor		
Thermal conductivity	34	– 55	W/m.K
Specific heat capacity	410	– 530	J/kg.K
Thermal expansion coefficient	10.5	– 13.5	µstrain/°C

Electrical properties
Electrical conductor or insulator?	Good conductor		
Electrical resistivity	*15	– 35	µohm.cm

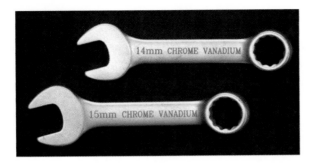

Low alloy chrome-molybdenum and chrome-vanadium steels are used for high-quality tools, bike frames, and automobile engine and transmission components.

Ecoproperties: material

Annual world production	1.1×10^9	– 1.2×10^9	tonne/yr
Reserves	78×10^9	– 79×10^9	tonne
Embodied energy, primary production	*32	– 38	MJ/kg
CO_2 footprint, primary production	*1.95	– 2.3	kg/kg
Water usage	*37	– 111	l/kg
Eco-indicator	100	– 120	millipoints/kg

Ecoproperties: processing

Casting energy	*3.8	– 4.6	MJ/kg
Casting CO_2 footprint	*0.23	– 0.28	kg/kg
Deformation processing energy	*3.2	– 3.9	MJ/kg
Deformation processing CO_2 footprint	*0.26	– 0.31	kg/kg

Recycling

Embodied energy, recycling	*9.0	– 11	MJ/kg
CO_2 footprint, recycling	*0.55	– *0.64	kg/kg
Recycle fraction in current supply	40	– 44	%

Typical uses. Springs, tools, ball bearings, rollers; crankshafts, gears, connecting rods, knives and scissors, pressure vessels.

Stainless steel

The material. Stainless steels are alloys of iron with chromium, nickel, and often four or five other elements. The alloying transmutes plain carbon steel that rusts and is prone to brittleness below room temperature into a material that does neither. Indeed, most stainless steels resist corrosion in most normal environments, and those that are *austenitic* (like AISI 302, 304, and 316) remain ductile to the lowest of temperatures.

Composition
Fe/ < 0.25C/16–30Cr/3.5–37Ni/ < 10Mn + Si,P,S (+N for 200 series).

General properties
Density	7600 – 8100	kg/m^3
Price	*7.7 – 8.5	USD/kg

Mechanical properties
Young's modulus	189 – 210	GPa
Yield strength (elastic limit)	170 – 1000	MPa
Tensile strength	480 – 2240	MPa
Elongation	5 – 70	%
Hardness—Vickers	130 – 570	HV
Fatigue strength at 10^7 cycles	*175 – 753	MPa
Fracture toughness	62 – 150	MPa.m$^{1/2}$

Thermal properties
Melting point	1370 – 1450	°C
Maximum service temperature	750 – 820	°C
Thermal conductor or insulator?	Poor conductor	
Thermal conductivity	12 – 24	W/m.K
Specific heat capacity	450 – 530	J/kg.K
Thermal expansion coefficient	13 – 20	µstrain/°C

Electrical properties
Electrical conductor or insulator?	Good conductor	
Electrical resistivity	64 – 107	µohm.cm

On the left: siemens toaster in brushed austenitic stainless steel (by Porsche Design). On the right, scissors in ferritic stainless steel; it is magnetic, whereas austenitic stainless is not.

Ecoproperties: material

Annual world production	30×10^6	– 3.1×10^6	tonne/yr
Reserves	*2.5×10^9	– 2.6×10^9	tonne
Embodied energy, primary production	*77	– 85	MJ/kg
CO_2 footprint, primary production	*4.7	– 5.4	kg/kg
Water usage	*112	– 336	l/kg
Eco-indicator	860	– 960	millipoints/kg

Ecoproperties: processing

Casting energy	*3.9	– 4.4	MJ/kg
Casting CO_2 footprint	*0.24	– 0.29	kg/kg
Deformation processing energy	*3.0	– 3.7	MJ/kg
Deformation processing CO_2 footprint	*0.26	– 0.31	kg/kg

Recycling

Embodied energy, recycling	*22	– 24	MJ/kg
CO_2 footprint, recycling	*1.4	– 1.5	kg/kg
Recycle fraction in current supply	35	– 40	%

Typical uses. Railway cars, trucks, trailers, food-processing equipment, sinks, stoves, cooking utensils, cutlery, flatware, scissors and knives, architectural metalwork, laundry equipment, chemical-processing equipment, jet-engine parts, surgical tools, furnace and boiler components, oil-burner parts, petroleum-processing equipment, dairy equipment, heat-treating equipment, automotive trim. Structural uses in corrosive environments, e.g., nuclear plants, ships, offshore oil installations, underwater cables, and pipes.

Cast iron, ductile (nodular)

The material. The foundations of modern industrial society are set, so to speak, in cast iron: it is the material that made the Industrial Revolution possible. Today it holds a second honor: that of being the cheapest of all engineering metals. Cast iron contains at least 2% carbon; most have 3–4%—and from 1–3% silicon. The carbon makes the iron very fluid when molten, allowing it to be cast to intricate shapes. There are five classes of cast iron: gray, white, ductile (or nodular), malleable, and alloy. The two that are most used are gray and ductile. This record is for ductile cast iron.

Composition
Fe/3.2–4.1% C/1.8–2.8% Si/ < 0.8% Mn/ <0.1% P/ <0.03% S

General properties

Density	7050	– 7250	kg/m^3
Price	*0.7	– 0.8	USD/kg

Mechanical properties

Young's modulus	165	– 180	GPa
Yield strength (elastic limit)	250	– 680	MPa
Tensile strength	410	– 830	MPa
Elongation	3	– 18	%
Hardness—Vickers	115	– 320	HV
Fatigue strength at 10^7 cycles	180	– 330	MPa
Fracture toughness	22	– 54	MPa.m$^{1/2}$

Thermal properties

Melting point	1130	– 1250	°C
Maximum service temperature	350	– 7450	°C
Thermal conductor or insulator?	Good conductor		
Thermal conductivity	29	– 44	W/m.K
Specific heat capacity	460	– 495	J/kg.K
Thermal expansion coefficient	10	– 12.5	µstrain/°C

Electrical properties

Electrical conductor or insulator?	Good conductor		
Electrical resistivity	49	– 56	µohm.cm

Ductile or malleable cast irons are used for heavily loaded parts such as gears and automotive suspension components.

Ecoproperties: material

Annual world production	1.1×10^9	– 1.2×10^9	tonne/yr
Reserves	78×10^9	– 79×10^9	tonne
Embodied energy, primary production	16	– 18	MJ/kg
CO_2 footprint, primary production	1.0	– 1.1	kg/kg
Water usage	*13	– 39	l/kg
Eco-indicator	38	– 42	millipoints/kg

Ecoproperties: processing

Casting energy	*3.2	– 3.7	MJ/kg
Casting CO_2 footprint	*0.19	– 0.22	kg/kg
Deformation processing energy	*2.5	– 3.1	MJ/kg
Deformation processing CO_2 footprint	*0.2	– 0.24	kg/kg

Recycling

Embodied energy, recycling	4.9	– 5.5	MJ/kg
CO_2 footprint, recycling	0.29	– 0.32	kg/kg
Recycle fraction in current supply	60	– 80	%

Typical uses. Brake discs and drums; bearings; camshafts; cylinder liners; piston rings; machine tool structural parts; engine blocks, gears, crankshafts; heavy-duty gear cases; pipe joints; pump casings; components in rock crushers.

12.3 Polymers

Polymers are the chemist's contribution to the materials world. The fact that most are derived from oil (a nonrenewable resource) and the difficulty of disposing of them at the end of their life (they don't easily degrade) has led to a view that polymers are environmental villains. There is some truth in this, but the present problems are soluble. Using oil to make polymers is a better primary use than just burning it for heat; the heat can still be recovered from the polymer at the end of its life. There are alternatives to oil; polymer feed stocks can be synthesized from agricultural products (notably starch and sugar, via methanol and ethanol). And thermoplastics—provided they are not contaminated—can be (and, to some extent, are) recycled.

Thermoplastics soften when heated and harden again to their original state when cooled. This allows them to be molded to complex shapes. Some are crystalline, some amorphous, some a mixture of both. Most accept coloring agents and fillers, and many can be blended to give a wide range of physical, visual, and tactile effects. Their sensitivity to sunlight is decreased by adding UV filters, and their flammability is decreased by adding flame retardants. The properties of thermoplastics can be controlled by chain length (measured by molecular weight), by degree of crystallinity, and by blending and plasticizing. As the molecular weight increases, the resin becomes stiffer, tougher, and more resistant to chemicals, but it is more difficult to mold. Crystalline polymers tend to have better chemical resistance, greater stability at high temperature, and better creep resistance than those that are amorphous. For transparency, the polymer must be amorphous; partial crystallinity gives translucency.

Thermosets. If you are a do-it-yourself type, you have Araldite in your toolbox—two tubes, one a sticky resin, the other an even stickier hardener. Mix and warm them and they react to give a stiff, strong, durable polymer, stuck to whatever it is put on. Araldite is an epoxy resin. It typifies *thermosets*: resins that polymerize when catalyzed and heated; when reheated they do not melt, they degrade. Polyurethane thermosets are produced in the highest volume; polyesters come second; phenolics (Bakelite), epoxies, and silicones follow, and, not surprisingly, the cost rises in the same order. Once shaped, thermosets cannot be reshaped. They cannot easily be recycled.

Elastomers were originally called "rubbers" because they could rub out pencil marks, but that is the least of their many remarkable and useful properties. Unlike any other class of solid, elastomers remember their shape when they are stretched and return to it when released. This allows *conformability*—hence their use for seals and gaskets. High-damping elastomers recover slowly; those with low damping snap back, returning the energy it took to stretch them—hence their use for springs, catapults, and bouncy things. Conformability gives elastomers high friction on rough surfaces, part of the reason (along with comfort) that they are used for pneumatic tires and footwear, their two largest markets.

Elastomers are thermosets; once cured, you can't remold them or recycle them, a major problem with car tires. Tricks can be used to make them behave in some ways like thermoplastics (TPOs—thermoplastic elastomers, of which EVA is an example). Blending or copolymerizing elastomer molecules with a thermoplastic like polypropylene (PP), if done properly, gives separated clumps of elastomer stuck together by a film of PP (Santoprene). The material behaves like an elastomer, but if heated so that the PP melts, it can be remolded and even recycled.

Profiles for 17 polymers follow, in the order in which they appear in Table 12.1.

Acrylonitrile butadiene styrene (ABS)

The material. Acrylonitrile butadiene styrene, or ABS, is tough, resilient, and easily molded. It is usually opaque, although some grades can now be transparent, and it can be given vivid colors. ABS-PVC alloys are tougher than standard ABS and, in self-extinguishing grades, are used for the casings of power tools.

Composition
$(CH_2—CH—C_6H_4)_n$

General properties

Density	1010	– 1210	kg/m^3
Price	2.3	– 2.6	USD/kg

Mechanical properties

Young's modulus	1.1	– 2.9	GPa
Yield strength (elastic limit)	18.5	– 51	MPa
Tensile strength	27.6	– 55.2	MPa
Elongation	1.5	– 100	%
Hardness—Vickers	5.6	– 15.3	HV
Fatigue strength at 10^7 cycles	11	– 22.1	MPa
Fracture toughness	1.19	– 4.29	$MPa.m^{1/2}$

Thermal properties

Glass temperature	88	– 128	°C
Maximum service temperature	62	– 77	°C
Thermal conductor or insulator?	Good insulator		
Thermal conductivity	0.188	– 0.335	W/m.K
Specific heat capacity	1390	– 1920	J/kg.K
Thermal expansion coefficient	84.6	– 234	μstrain/°C

Electrical properties

Electrical conductor or insulator?	Good insulator		
Electrical resistivity	3.3×10^{21}	– 3×10^{22}	μohm.cm
Dielectric constant	2.8	– 3.2	
Dissipation factor	0.003	– 0.007	
Dielectric strength	13.8	– 21.7 10^6	V/m

ABS allows detailed moldings, accepts color well, and is nontoxic and tough.

Ecoproperties: material

Annual world production	*5.6×10^6	– 5.7×10^6	tonne/yr
Reserves	*1.48×10^8	– 1.5×10^8	tonne
Embodied energy, primary production	*91	– 102	MJ/kg *96.5*
CO_2 footprint, primary production	*3.3	– 3.6	kg/kg *3.45*
Water usage	*108	– 324	l/kg
Eco-indicator	380	– 420	millipoints/kg

Ecoproperties: processing

Polymer molding energy	*10	– 12	MJ/kg *11*
Polymer molding CO_2 footprint	*0.8	– 0.96	kg/kg *.88*
Polymer extrusion energy	*3.2	– 4.6	MJ/kg *3.9 .34*
Polymer extrusion CO_2 footprint	*0.31	– 0.37	kg/kg *.34*

≈ 6.1

≈ 141

Recycling

Embodied energy, recycling	*38	– 43	MJ/kg *41.5*
CO_2 footprint, recycling	*1.39	– 1.5	kg/kg *1.45*
Recycle fraction in current supply	0.5	– 1	%
Recycle mark			

Other

Typical uses. Safety helmets; camper tops; automotive instrument panels and other interior components; pipe fittings; home-security devices and housings for small appliances; communications equipment; business machines; plumbing hardware; automobile grilles; wheel covers; mirror housings; refrigerator liners; luggage shells; tote trays; mower shrouds; boat hulls; large components for recreational vehicles; weather seals; glass beading; refrigerator breaker strips; conduit; pipe for drain-waste-vent (DWV) systems.

Polyamides (Nylons, PA)

The material. Back in 1945, the war in Europe just ended, the two most prized luxuries were cigarettes and stockings made of nylon. Nylon (PA) can be drawn to fibers as fine as silk and was widely used as a substitute for it. Today, newer fibers have eroded its dominance in garment design, but nylon-fiber ropes and nylon as reinforcement for rubber (in car tires) and other polymers (PTFE, for roofs) remain important. Nylon is used in product design for tough casings, frames and handles, and, reinforced with glass, as bearings gears and other load-bearing parts. There are many grades (Nylon 6, Nylon 66, Nylon 11, etc.), each with slightly different properties.

Composition
$(NH(CH_2)_5CO)_n$

General properties

Density	1120	– 1140	kg/m^3
Price	*3.55	– 3.91	USD/kg

Mechanical properties

Young's modulus	2.62	– 3.2	GPa
Yield strength (elastic limit)	50	– 94.8	MPa
Tensile strength	90	– 165	MPa
Elongation	30	– 100	%
Hardness—Vickers	25.8	– 28.4	HV
Fatigue strength at 10^7 cycles	*36	– 66	MPa
Fracture toughness	*2.2	– 5.6	MPa.m$^{1/2}$

Thermal properties

Melting point	210	– 220	°C
Maximum service temperature	110	– 140	°C
Thermal conductor or insulator?	Good insulator		
Thermal conductivity	0.23	– 0.25	W/m.K
Specific heat capacity	*1600	– 1660	J/kg.K
Thermal expansion coefficient	144	– 149	µstrain/°C

Electrical properties

Electrical conductor or insulator?	Good insulator		
Electrical resistivity	*1.5×10^{19}	– 1.4×10^{20}	µohm.cm
Dielectric constant	3.7	– 3.9	
Dissipation factor	0.014	– 0.03	
Dielectric strength	15.1	– 16.4×10^6	V/m

Polyamides are tough, wear and corrosion resistant, and can be colored.

Ecoproperties: material

Annual world production	3.7×10^6	– 3.8×10^6	tonne/yr
Reserves	$^*9.2 \times 10^8$	– 9.3×10^8	tonne
Embodied energy, primary production	121	– 135	MJ/kg
CO_2 footprint, primary production	5.5	– 5.6	kg/kg
Water usage	^{*136}	– 408	l/kg
Eco-indicator	600	– 660	millipoints/kg

Ecoproperties: processing

Polymer molding energy	$^*9.2$	– 10	MJ/kg
Polymer molding CO_2 footprint	$^*0.74$	– 0.81	kg/kg
Polymer extrusion energy	$^*3.6$	– 3.9	MJ/kg
Polymer extrusion CO_2 footprint	$^*0.29$	– 0.32	kg/kg

Recycling

Embodied energy, recycling	50.8	– 56.7	MJ/kg
CO_2 footprint, recycling	2.31	– 2.35	kg/kg
Recycle fraction in current supply	$^*0.5$	– 1	%
Recycle mark			

Other

Typical uses. Light duty gears, bushings, sprockets and bearings; electrical equipment housings, lenses, containers, tanks, tubing, furniture casters, plumbing connections, bicycle wheel covers, ketchup bottles, chairs, toothbrush bristles, handles, bearings, food packaging. Nylons are used as hot-melt adhesives for book bindings; as fibers—ropes, fishing line, carpeting, car upholstery and stockings; as aramid fibers—cables, ropes, protective clothing, air filtration bags and electrical insulation.

Polypropylene (PP)

The material. Polypropylene, or PP, first produced commercially in 1958, is the younger brother of polyethylene, a very similar molecule with similar price, processing methods, and application. Like PE it is produced in very large quantities (more than 30 million tons per year in 2000), growing at nearly 10% per year, and like PE its molecule lengths and side branches can be tailored by clever catalysis, giving precise control of impact strength, and of the properties that influence molding and drawing. In its pure form polypropylene is flammable and degrades in sunlight. Fire retardants make it slow to burn and stabilizers give it extreme stability, both to UV radiation and to fresh and saltwater and most aqueous solutions.

Composition
$(CH_2—CH(CH_3))_n$

General properties

Density	890	– 910	kg/m^3
Price	*2.1	– 2.35	USD/kg

Mechanical properties

Young's modulus	0.9	– 1.55	GPa
Yield strength (elastic limit)	21	– 37	MPa
Tensile strength	28	– 41	MPa
Elongation	100	– 600	%
Hardness-Vickers	6.2	– 11	HV
Fatigue strength at 10^7 cycles	11	– 17	MPa
Fracture toughness	3	– 4.5	MPa.m$^{1/2}$

Thermal properties

Melting point	150	– 175	°C
Maximum service temperature	100	– 115	°C
Thermal conductor or insulator?	Good insulator		
Thermal conductivity	0.11	– 0.17	W/m.K
Specific heat capacity	1870	– 1960	J/kg.K
Thermal expansion coefficient	122	– 180	μstrain/°C

Electrical properties

Electrical conductor or insulator?	Good insulator		
Electrical resistivity	3.3×10^{22}	– 3×10^{23}	μohm.cm
Dielectric constant	2.1	– 2.3	
Dissipation factor	3×10^{-4}	– 7×10^{-4}	
Dielectric strength	22.7	– 24.6×10^6	V/m

Polypropylene is widely used in household products.

Ecoproperties: material

Annual world production	43×10^6	–	44×10^6	tonne/yr
Reserves	$^{\star}1.2 \times 10^9$	–	1.3×10^9	tonne
Embodied energy, primary production	85	–	1.10	MJ/kg
CO_2 footprint, primary production	2.6	–	2.8	kg/kg
Water usage	$^{\star}50$	–	150	l/kg

Ecoproperties: processing

Polymer molding energy	$^{\star}8.2$	–	9	MJ/kg
Polymer molding CO_2 footprint	$^{\star}0.65$	–	0.72	kg/kg
Polymer extrusion energy	$^{\star}3.2$	–	3.5	MJ/kg
Polymer extrusion CO_2 footprint	$^{\star}0.25$	–	0.28	kg/kg

Recycling

Embodied energy, recycling	36	–	44	MJ/kg
CO_2 footprint, recycling	1.1	–	1.2	kg/kg
Recycle fraction in current supply	5.1	–	6	%
Recycle mark				

PP

Typical uses. Ropes, general polymer engineering, automobile air ducting, parcel shelving and air-cleaners, garden furniture, washing machine tank, wet-cell battery cases, pipes and pipe fittings, beer bottle crates, chair shells, capacitor dielectrics, cable insulation, kitchen kettles, car bumpers, shatter proof glasses, crates, suitcases, artificial turf, thermal underwear.

Polyethylene (PE)

The material. Polyethylene, $(—CH_2—)_n$, first synthesized in 1933, looks like the simplest of molecules, but the number of ways in which the—CH_2 units can be linked is large. It is the first of the polyolefins, the bulk thermoplastic polymers that account for a dominant fraction of all polymer consumption. Polyethylene is inert, and extremely resistant to fresh and saltwater, food, and most water-based solutions. For this reason it is widely used in household products, food containers, and chopping boards. Polyethylene is cheap and particularly easy to mold and fabricate. It accepts a wide range of colors, can be transparent, translucent or opaque, has a pleasant, slightly waxy feel, can be textured or metal coated, but is difficult to print on.

Composition
$(—CH_2—CH_2—)_n$

General properties

Density	939	– 960	kg/m^3
Price	1.85	– 2	USD/kg

Mechanical properties

Young's modulus	0.62	– 0.86	GPa
Yield strength (elastic limit)	18	– 29	MPa
Tensile strength	21	– 45	MPa
Elongation	200	– 800	%
Hardness—Vickers	5.4	– 8.7	HV
Fatigue strength at 10^7 cycles	21	– 23	MPa
Fracture toughness	*1.4	– 1.7	MPa.m$^{1/2}$

Thermal properties

Melting point	125	– 132	°C
Maximum service temperature	*90	– 110	°C
Thermal conductor or insulator?	Good insulator		
Thermal conductivity	0.4	– 0.44	W/m.K
Specific heat capacity	*1810	– 1880	J/kg.K
Thermal expansion coefficient	126	– 198	μstrain/°C

Electrical properties

Electrical conductor or insulator?	Good insulator		
Electrical resistivity	3.3×10^{22}	– 3×10^{24}	μohm.cm
Dielectric constant	2.2	– 2.4	
Dissipation factor	*3×10^{-4}	– 6×10^{-4}	
Dielectric strength	17.7	– 19.7 10^6	V/m

PE is widely used for containers and packaging.

Ecoproperties: material

Annual world production	68×10^6	– 69×10^6	tonne/yr
Reserves	$*1.7 \times 10^9$		tonne
Embodied energy, primary production	77	– 85	MJ/kg
CO_2 footprint, primary production	2	– 2.2	kg/kg
Water usage	*38	– 1.1e2	l/kg
Eco-indicator	310	– 350	millipoints/kg

Ecoproperties: processing

Polymer molding energy	*6.1	– 6.8	MJ/kg
Polymer molding CO_2 footprint	*0.49	– 0.54	kg/kg
Polymer extrusion energy	*2.4	– 2.7	MJ/kg
Polymer extrusion CO_2 footprint	*0.19	– 0.21	kg/kg

Recycling

Embodied energy, recycling	32	– 36	MJ/kg
CO_2 footprint, recycling	0.82	– 0.91	kg/kg
Recycle fraction in current supply	7.5	– 9.5	%
Recycle mark			

HDPE

LDPE

Typical uses. Oil container, street bollards, milk bottles, toys, beer crate, food packaging, shrink wrap, squeeze tubes, disposable clothing, plastic bags, paper coatings, cable insulation, artificial joints, and as fibers—low-cost ropes and packing tape reinforcement.

Polycarbonate (PC)

The material. PC is one of the "engineering" thermoplastics, meaning that they have better mechanical properties than the cheaper "commodity" polymers. The benzene ring and the —OCOO-carbonate group combine in pure PC to give it its unique characteristics of optical transparency and good toughness and rigidity, even at relatively high temperatures. These properties make PC a good choice for applications such as compact disks, safety hard hats, and housings for power tools.

Composition
$(O—(C_6H_4)—C(CH_3)_2—(C_6H_4)—CO)_n$

General properties

Density	1140	– 1210	kg/m^3
Price	3.52	– 3.87	USD/kg

Mechanical properties

Young's modulus	2	– 2.44	GPa
Yield strength (elastic limit)	59	– 70	MPa
Tensile strength	60	– 72.4	MPa
Compressive strength	69	– 86.9	MPa
Elongation	70	– 150	%
Hardness—Vickers	17.7	– 21.7	HV
Fatigue strength at 10^7 cycles	22.1	– 30.8	MPa
Fracture toughness	2.1	– 4.6	MPa.m$^{1/2}$

Thermal properties

Glass temperature	142	– 205	°C
Maximum service temperature	101	– 144	°C
Thermal conductor or insulator?	Good insulator		
Thermal conductivity	0.189	– 0.218	W/m.K
Specific heat capacity	1530	– 1630	J/kg.K
Thermal expansion coefficient	120	– 137	μstrain/°C

Electrical properties

Electrical conductor or insulator?	Good insulator		
Electrical resistivity	1×10^{20}	– 1×10^{22}	μohm.cm
Dielectric constant	3.1	– 3.3	
Dissipation factor	8×10^{-4}	– 0.0011	
Dielectric strength	15.7	– 19.2×10^6	V/m

Polycarbonate is tough and impact resistant: hence its use in hard hats and helmets, transparent roofing, and riot shields.

Ecoproperties: material

Embodied energy, primary production	105	– 116	MJ/kg	5.65
CO_2 footprint, primary production	5.4	– 5.9	kg/kg	
Water usage	*142	– 425	l/kg	
Eco-indicator	480	– 540	millipoints/kg	

(handwritten: 110.5)

Ecoproperties: processing

Polymer molding energy	*10.2	– 11.2	MJ/kg	.860
Polymer molding CO_2	*0.814	– 0.898	kg/kg	
Polymer extrusion energy	*4.4	– 4.99	MJ/kg	
Polymer extrusion CO_2	*0.352	– 0.399	kg/kg	.375

(handwritten: 10.7   4.7)

Recycling

Embodied energy, recycling	44.1	– 48.7	MJ/kg	2.385
CO_2 footprint, recycling	2.27	– 2.48	kg/kg	
Recycle fraction in current supply	*0.5	– 1	%	
Recycle mark				

(handwritten: 46.1)

(handwritten: .2.76   82   3.62)

Other

Typical uses. Safety shields and goggles; lenses; glazing panels; business machine housing; instrument casings; lighting fittings; safety helmets; electrical switchgear; laminated sheet for bulletproof glazing; twin-walled sheets for glazing; kitchenware and tableware; microwave cookware; medical (sterilizable) components.

Polyethylene terephthalate (PET)

The material. The name *polyester* derives from a combination of *Polymerization* and *esterification*. Saturated polyesters are thermoplastic; examples are PET and PBT, which have good mechanical properties to temperatures as high as 175°C. PET is crystal clear and impervious to water and CO_2, but a little oxygen does get through. It is tough, strong, and easy to shape, join, and sterilize, allowing reuse. Unsaturated polyesters are thermosets; they are used as the matrix material in glass fiber/polyester composites.

Composition
$(CO—(C_6H_4)—CO—O—(CH_2)_2—O)_n$

General properties

Density	1290	– 1400	kg/m^3
Price	*1.63	– 1.79	USD/kg

Mechanical properties

Young's modulus	2.76	– 4.14	GPa
Yield strength (elastic limit)	56.5	– 62.3	MPa
Tensile strength	48.3	– 72.4	MPa
Compressive strength	62.2	– 68.5	MPa
Elongation	30	– 300	%
Hardness—Vickers	17	– 18.7	HV
Fatigue strength at 10^7 cycles	*19.3	– 29	MPa
Fracture toughness	4.5	– 5.5	MPa.m$^{1/2}$

Thermal properties

Melting point	212	– 265	°C
Glass temperature	67.9	– 79.9	°C
Maximum service temperature	66.9	– 86.9	°C
Thermal conductor or insulator?	Good insulator		
Thermal conductivity	0.138	– 0.151	W/m.K
Specific heat capacity	*1420	– 1470	J/kg.K
Thermal expansion coefficient	115	– 119	μstrain/°C

Electrical properties

Electrical conductor or insulator?	Good insulator		
Electrical resistivity	3.3×10^{20}	– 3.0×10^{21}	μohm.cm
Dielectric constant	3.5	– 3.7	
Dissipation factor	*0.003	– 0.007	
Dielectric strength	16.5	– 21.7 10^6	V/m

PET drinks containers, pressurized and unpressurized.

Ecoproperties: material

Annual world production	9×10^6	9.2×10^6	tonne/yr
Reserves	$^*2.5 \times 10^8$	2.6×10^8	tonne
Embodied energy, primary production	79.6	88	MJ/kg
CO_2 footprint, primary production	2.21	2.45	kg/kg
Water usage	$^*14.7$	44.2	l/kg
Eco-indicator	360	400	millipoints/kg

84 (handwritten)

2.33 (handwritten)

Ecoproperties: processing

Polymer molding energy	$^*9.36$	10.3	MJ/kg
Polymer molding CO_2	$^*0.749$	0.826	kg/kg
Polymer extrusion energy	$^*3.63$	4	MJ/kg
Polymer extrusion CO_2	$^*0.29$	0.32	kg/kg

9.75 (handwritten)
0.79 (handwritten)
3.85 (handwritten)
0.31 (handwritten)

≈4.4 (handwritten, circled)

Recycling

Embodied energy, recycling	33.4	37	MJ/kg
CO_2 footprint, recycling	0.928	1.03	kg/kg
Recycle fraction in current supply	20	22	%
Recycle mark			

-97 (handwritten)

≈131 (handwritten, circled)

PET

97.6 (handwritten)

Typical uses. Electrical fittings and connectors; blow molded bottles; packaging film; film; photographic and X-ray film; audio/visual tapes; industrial strapping; capacitor film; drawing office transparencies; fibers. Decorative film, metallized balloons, photography tape, videotape, carbonated drink containers, ovenproof cookware, windsurfing sails, credit cards.

Polyvinylchloride (tpPVC)

The material. PVC (vinyl) is one of the cheapest, most versatile, and, with polyethylene, the most widely used of polymers and epitomizes their multi-faceted character. In its pure form—as a thermoplastic, tpPVC—it is rigid and not very tough; its low price makes it a cost-effective engineering plastic where extremes of service are not encountered. Incorporating plasticizers creates flexible PVC, elPVC, a material with leather- or rubberlike properties and used a substitute for both. By contrast, reinforcement with glass fibers gives a material that is sufficiently stiff, strong, and tough to be used for roofs, flooring, and building panels.

Composition
$(CH_2CHCl)_n$

General properties

Density	1300	– 1580	kg/m^3
Price	1.52	– 1.67	USD/kg

Mechanical properties

Young's modulus	2.14	– 4.14	GPa
Yield strength (elastic limit)	35.4	– 52.1	MPa
Tensile strength	40.7	– 65.1	MPa
Compressive strength	42.5	– 89.6	MPa
Elongation	11.9	– 80	%
Hardness—Vickers	10.6	– 15.6	HV
Fatigue strength at 10^7 cycles	16.2	– 26.1	MPa
Fracture toughness	1.46	– 5.12	MPa.m$^{1/2}$

Thermal properties

Glass temperature	74.9	– 105	°C
Maximum service temperature	60	– 70	°C
Thermal conductor or insulator?	Good insulator		
Thermal conductivity	0.147	– 0.293	W/m.K
Specific heat capacity	1360	– 1440	J/kg.K
Thermal expansion coefficient	100	– 150	μstrain/°C

Electrical properties

Electrical conductor or insulator?	Good insulator		
Electrical resistivity	1×10^{20}	– 1×10^{22}	μohm.cm
Dielectric constant	3.1	– 4.4	
Dissipation factor	0.03	– 0.1	
Dielectric strength	13.8	– 19.7 10^6	V/m

These boat fenders illustrate that PVC is tough, weather resistant, and easy to form and color.

Ecoproperties: material

Annual world production	4.9×10^7	– 5.1×10^7	tonne/yr
Reserves	*1.38×10^9	– 1.4×10^9	tonne
Embodied energy, primary production	68	– 95	MJ/kg
CO_2 footprint, primary production	2.2	– 2.6	kg/kg
Water usage	*18.9	– 56.7	l/kg
Eco-indicator	260	– 280	millipoints/kg

Ecoproperties: processing

Polymer molding energy	*9.57	– 10.6	MJ/kg
Polymer molding CO_2	*0.766	– 0.844	kg/kg
Polymer extrusion energy	*3.41	– 3.75	MJ/kg
Polymer extrusion CO_2	*0.272	– 0.3	kg/kg

Recycling

Embodied energy, recycling	28.6	– 39.9	MJ/kg
CO_2 footprint, recycling	0.924	– 1.09	kg/kg
Recycle fraction in current supply	0.5	– 1	%
Recycle mark			

PVC

Typical uses. tpPVC: pipes, fittings, profiles, road signs, cosmetic packaging, canoes, garden hoses, vinyl flooring, windows and cladding, vinyl records, dolls, medical tubes. elPVC: artificial leather, wire insulation, film, sheet, fabric, car upholstery.

Polystyrene (PS)

The material. Polystyrene is an optically clear, cheap, easily molded polymer, familiar as the standard "jewel" CD case. In its simplest form PS is brittle. Its mechanical properties are dramatically improved by blending with polybutadiene but with a loss of optical transparency. High-impact PS (10% polybutadiene) is much stronger, even at low temperatures (meaning strength down to −12°C). The single largest use of PS is a foam packaging.

Composition
$(CH(C_6H_5)—CH_2)n$

General properties

Density	1040	– 1050	kg/m^3
Price	1.81	– 1.99	USD/kg

Mechanical properties

Young's modulus	1.2	– 2.6	GPa
Yield strength (elastic limit)	28.7	– 56.2	MPa
Tensile strength	35.9	– 56.5	MPa
Compressive strength	31.6	– 61.8	MPa
Elongation	1.2	– 3.6	%
Hardness—Vickers	8.6	– 16.9	HV
Fatigue strength at 10^7 cycles	14.4	– 23	MPa
Fracture toughness	0.7	– 1.1	$MPa.m^{1/2}$

Thermal properties

Glass temperature	73.9	– 110	°C
Maximum service temperature	76.9	– 103	°C
Thermal conductor or insulator?	Good insulator		
Thermal conductivity	0.121	– 0.131	W/m.K
Specific heat capacity	1690	– 1760	J/kg.K
Thermal expansion coefficient	90	– 153	μstrain/°C

Electrical properties

Electrical conductor or insulator?	Good insulator		
Electrical resistivity	1×10^{25}	– 1×10^{27}	μohm.cm
Dielectric constant	3	– 3.2	
Dissipation factor	0.001	– 0.003	
Dielectric strength	19.7	– 22.6 10^6	V/m

Polystyrene is water-clear, easily formed, and cheap.

Ecoproperties: material

Annual world production	1.2×10^7	– 1.22×10^7	tonne/yr
Reserves	*3×10^8	– 3.1×10^8	tonne
Embodied energy, primary production	86	– 99	MJ/kg
CO_2 footprint, primary production	2.7	– 3	kg/kg
Water usage	*108	– 323	l/kg
Eco-indicator	340	– 380	millipoints/kg

Ecoproperties: processing

Polymer molding energy	*10.3	– 11.3	MJ/kg
Polymer molding CO_2	*0.82	– 0.90	kg/kg
Polymer extrusion energy	*3.97	– 4.37	MJ/kg
Polymer extrusion CO_2	*0.31	– 0.35	kg/kg

Recycling

Embodied energy, recycling	36.1	– 41.6	MJ/kg
CO_2 footprint, recycling	1.1	– 1.2	kg/kg
Recycle fraction in current supply	2.1	– 3	%
Recycle mark			

PS

Typical uses. Toys; light diffusers; lenses and mirrors; beakers; cutlery; general household appliances; video/audio cassette cases; electronic housings; refrigerator liners.

Polylactide (PLA)

The material. Polylactide (PLA) is a biodegradable thermoplastic derived from natural lactic acid from corn, maize, or milk. It resembles clear polystyrene and provides good aesthetics (gloss and clarity), but it is stiff and brittle and needs modification using plasticizers for most practical applications. It can be processed like most thermoplastics into fibers, films, thermoformed, or injection molded.

General properties

Density	1210	– 1250	kg/m^3
Price	*2	– 4	USD/kg

Mechanical properties

Young's modulus	3.45	– 3.83	GPa
Yield strength (elastic limit)	48	– 60	MPa
Tensile strength	48	– 60	MPa
Compressive strength	48	– 60	MPa
Elongation	5	– 7	%
Hardness—Vickers	*14	– 18	HV
Fatigue strength at 10^7 cycles	*14	– 18	MPa
Fracture toughness	*0.7	– 1.1	MPa.m$^{1/2}$

Thermal properties

Melting point	160	– 177	°C
Glass temperature	56	– 58	°C
Maximum service temperature	70	– 80	°C
Thermal conductor or insulator?	Good insulator		
Thermal conductivity	0.12	– 0.13	W/m.K
Specific heat capacity	1180	– 1210	J/kg.K
Thermal expansion coefficient	*126	– 145	μstrain/°C

Electrical properties

Electrical conductor or insulator?	Good insulator		
Electrical resistivity	*1×10^{17}	– 1×10^{19}	μohm.cm
Dielectric constant	*3.5	– 5	
Dissipation factor	*0.02	– 0.07	
Dielectric strength	12	– 16 10^6	V/m

Cargill Dow polylactide food packaging.

Ecoproperties: material

Embodied energy, primary production	52	–	54	MJ/kg
CO_2 footprint, primary production	*2.3	–	2.43	kg/kg
Water usage	*100	–	300	l/kg

Ecoproperties: processing

Polymer molding energy	*8.61	–	9.49	MJ/kg
Polymer molding CO_2	*0.68	–	0.75	kg/kg
Polymer extrusion energy	*3.35	–	3.69	MJ/kg
Polymer extrusion CO_2	*0.26	–	0.29	kg/kg

Recycling

Embodied energy, recycling	21.8	–	22.7	MJ/kg
CO_2 footprint, recycling	*0.96	–	1.02	kg/kg
Recycle fraction in current supply	*0.5	–	1	%
Recycle mark				

Other

Typical uses. Food packaging, plastic bags, plant pots, diapers, bottles, cold drink cups, sheet and film.

Polyhydroxyalkanoates (PHA, PHB)

The material. PHAs are linear polyesters produced in nature by bacterial fermentation of sugar or lipids derived from soybean oil, corn oil, or palm oil. They are fully biodegradable. More than 100 different monomers can be combined within this family to give materials with a wide range of properties, from stiff and brittle thermoplastics to flexible elastomers. The most common type of PHA is PHB (poly-3-hydroxybutyrate), with properties similar to those of PP, though it is stiffer and more brittle. The following data is for PHB.

Composition
$(CH(CH_3)—CH_2—CO—O)_n$

General properties

Density	1230	– 1250	kg/m^3
Price	3.2	– 4	USD/kg

Mechanical properties

Young's modulus	0.8	– 4	GPa
Yield strength (elastic limit)	35	– 40	MPa
Tensile strength	35	– 40	MPa
Compressive strength	*40	– 45	MPa
Elongation	6	– 25	%
Hardness—Vickers	*11	– 13	HV
Fatigue strength at 10^7 cycles	*12	– 17	MPa
Fracture toughness	*0.7	– 1.2	MPa.m$^{1/2}$

Thermal properties

Melting point	115	– 175	°C
Glass temperature	4	– 15	°C
Maximum service temperature	*60	– 80	°C
Thermal conductor or insulator?	Good insulator		
Thermal conductivity	*0.13	– 0.23	W/m.K
Specific heat capacity	*1400	– 1600	J/kg.K
Thermal expansion coefficient	*180	– 240	μstrain/°C

Electrical properties

Electrical conductor or insulator?	Good insulator		
Electrical resistivity	*1×10^{16}	– 1×10^{18}	μohm.cm
Dielectric constant	*3	– 5	
Dissipation factor	*0.05	– 0.15	
Dielectric strength	12	– 16 10^6	V/m

PHB containers. (Kumar and Minocha, Trangenic Plant Research, Harwood Publishers.)

Ecoproperties: material

Embodied energy, primary production	50	–	59	MJ/kg
CO_2 footprint, primary production	*2.25	–	2.66	kg/kg
Water usage	*100	–	300	l/kg

Ecoproperties: processing

Polymer molding energy	*8.24	–	9.09	MJ/kg
Polymer molding CO_2	*0.65	–	0.72	kg/kg
Polymer extrusion energy	*3.21	–	3.54	MJ/kg
Polymer extrusion CO_2	*0.25	–	0.28	kg/kg

Recycling

Embodied energy, recycling	21	–	24.8	MJ/kg
CO_2 footprint, recycling	*0.94	–	1.12	kg/kg
Recycle fraction in current supply	0.5	–	1	%
Recycle mark				

Other

Typical uses. Packaging, containers, bottles.

Epoxies

The material. Epoxies are thermosetting polymers with excellent mechanical, electrical, and adhesive properties and good resistance to heat and chemical attack. They are used for adhesives (Araldite), surface coatings, and, when filled with other materials such as glass or carbon fibers, as matrix resins in composite materials. Typically, as adhesives, epoxies are used for high-strength bonding of dissimilar materials; as coatings, they are used to encapsulate electrical coils and electronic components; when filled, they are used for tooling fixtures for low-volume molding of thermoplastics.

Composition
$(O—C_6H_4—CH_3—C—CH_3—C_6H_4)_n$

General properties

Density	1110	– 1400	kg/m^3
Price	2.2	– 2.47	USD/kg

Mechanical properties

Young's modulus	2.35	– 3.08	GPa
Yield strength (elastic limit)	36	– 71.7	MPa
Tensile strength	45	– 89.6	MPa
Compressive strength	39.6	– 78.8	MPa
Elongation	2	– 10	%
Hardness—Vickers	10.8	– 21.5	HV
Fatigue strength at 10^7 cycles	*22.1	– 35	MPa
Fracture toughness	0.4	– 2.22	MPa.m$^{1/2}$

Thermal properties

Glass temperature	66.9	– 167	°C
Maximum service temperature	140	– 180	°C
Thermal conductor or insulator?	Good insulator		
Thermal conductivity	0.18	– 0.5	W/m.K
Specific heat capacity	1490	– 2000	J/kg.K
Thermal expansion coefficient	58	– 117	µstrain/°C

Electrical properties

Electrical conductor or insulator?	Good insulator		
Electrical resistivity	1×10^{20}	– 6×19^{21}	µohm.cm
Dielectric constant	3.4	– 5.7	
Dissipation factor	7×10^{-4}	– 0.015	
Dielectric strength	11.8	– 19.7 10^6	V/m

Epoxy paints are exceptionally stable and protective and take color well. Epoxies are used at the matrix of high-performance composite and as high-strength adhesives.

Ecoproperties: material

Annual world production	1.2×10^5	$-$ 1.25×10^5	tonne/yr
Reserves	*3.7×10^6	$-$ 3.8×10^6	tonne
Embodied energy, primary production	105	$-$ 130	MJ/kg
CO_2 footprint, primary production	4.22	$-$ 4.56	kg/kg
Water usage	*107	$-$ 322	l/kg

Ecoproperties: processing

Polymer molding energy	*10.3	$-$ 12.5	MJ/kg
Polymer molding CO_2	*0.825	$-$ 1.0	kg/kg

Recycling

Recycle fraction in current supply	0.5	$-$ 1	%

Typical uses. Pure epoxy molding compounds: the encapsulation of electrical coils and electronics components; epoxy resins in laminates: pultruded rods, girder stock, special tooling fixtures, mechanical components such as gears; adhesives, often for high-strength bonding of dissimilar materials; patterns and molds for shaping thermoplastics.

Polyester

The material. Polyesters can be thermosets, thermoplastics, or elastomers. The unsaturated polyester resins are thermosets. Most polyester thermosets are used in glass fiber/polyester composites. They are less stiff and strong than epoxies, but they are considerably cheaper. This record is for thermosetting polyester. It cannot be recycled.

Composition
$(OOC—C_6H_4—COO—C_6H_{10})_n$

General properties

Density	1040	– 1400	kg/m^3
Price	*4.07	– 4.47	USD/kg

Mechanical properties

Young's modulus	2.07	– 4.41	GPa
Yield strength (elastic limit)	*33	– 40	MPa
Tensile strength	41.4	– 89.6	MPa
Compressive strength	*36.3	– 44	MPa
Elongation	2	– 2.6	%
Hardness—Vickers	9.9	– 21.5	HV
Fatigue strength at 10^7 cycles	*16.6	– 35.8	MPa
Fracture toughness	*1.09	– 1.69	MPa.m$^{1/2}$

Thermal properties

Glass temperature	147	– 207	°C
Maximum service temperature	130	– 150	°C
Thermal conductor or insulator?	Good insulator		
Thermal conductivity	*0.287	– 0.299	W/m.K
Specific heat capacity	*1510	– 1570	J/kg.K
Thermal expansion coefficient	99	– 180	μstrain/°C

Electrical properties

Electrical conductor or insulator?	Good insulator		
Electrical resistivity	3.3×10^{18}	– 3×10^{19}	μohm.cm
Dielectric constant	2.8	– 3.3	
Dissipation factor	0.001	– 0.03	
Dielectric strength	15	– 19.7 10^6	V/m

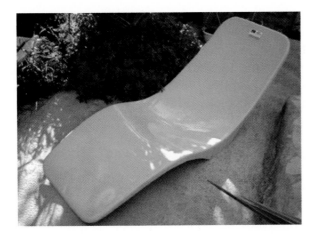

Thermosetting polyester is used as the matrix of this glass-reinforced deck chair.

Ecoproperties: material

Annual world production	4×10^7	– 4.05×10^7	tonne/yr
Reserves	$^*1 \times 10^9$	– 1.01×10^9	tonne
Embodied energy, primary production	^{*84}	– 93	MJ/kg
CO_2 footprint, primary production	$^*2.7$	– 3	kg/kg
Water usage	$^*88.1$	– 264	l/kg

Ecoproperties: processing

Polymer molding energy	$^*11.4$	– 12.8	MJ/kg
Polymer molding CO_2	$^*0.91$	– 1.0	kg/kg

Recycling

Recycle fraction in current supply	0.1	%

Typical uses. Laminated structures; surface gel coatings; liquid castings; furniture products; bowling balls; simulated marble; sewer pipe gaskets; pistol grips; television tube implosion barriers; boats; truck cabs; concrete forms; lamp housings; skylights; fishing rods.

Phenolics

The material. Bakelite, commercialized in 1909, triggered a revolution in product design. It was stiff, fairly strong, could (to a muted degree) be colored, and, above all, was easy to mold. Earlier products that were hand-crafted from woods, metals, or exotics such as ivory, could now be molded quickly and cheaply. At one time the production of phenolics exceeded that of PE, PS, and PVC combined. Now, although the ration has changed, phenolics still have a unique value. They are stiff, are chemically stable, have good electrical properties, are fire-resistant, and are easy to mold—and they are cheap. Thermosetting phenolics are recyclable but by a different means than that for thermoplastics. Molded phenolic, ground into a fine powder, can be added to the raw material stream. Four percent to 12% ground phenolic does not degrade properties.

General properties

Density	1240	– 1320	kg/m^3
Price	1.65	– 1.87	USD/kg

Mechanical properties

Young's modulus	2.76	– 4.83	GPa
Yield strength (elastic limit)	*27.6	– 49.7	MPa
Tensile strength	34.5	– 62.1	MPa
Compressive strength	*30.4	– 54.6	MPa
Elongation	1.5	– 2	%
Hardness—Vickers	8.3	– 14.9	HV
Fatigue strength at 10^7 cycles	*13.8	– 24.8	MPa
Fracture toughness	*0.787	– 1.21	MPa.m$^{1/2}$

Thermal properties

Glass temperature	167	– 267	°C
Maximum service temperature	*200	– 230	°C
Thermal conductor or insulator?	Good insulator		
Thermal conductivity	0.14	– 0.15	W/m.K
Specific heat capacity	*1470	– 1530	J/kg.K
Thermal expansion coefficient	120	– 125	µstrain/°C

Electrical properties

Electrical conductor or insulator?	Good insulator		
Electrical resistivity	3.3×10^{18}	– 3×10^{19}	µohm.cm
Dielectric constant	*4	– 6	
Dissipation factor	*0.005	– 0.01	
Dielectric strength	9.84	– 15.7 10^6	V/m

Phenolics are good insulators and resist heat and chemical attack exceptionally well, making them a good choice for electrical switchgear like this distributor cap.

Ecoproperties: material

Annual world production	1.0×10^7	– 1.1×10^7	tonne/yr
Reserves	*2.5×10^8	– 2.5×10^8	tonne
Embodied energy, primary production	*85.9	– 95	MJ/kg
CO_2 footprint, primary production	*2.83	– 3.12	kg/kg
Water usage	*94	– 282	l/kg

Ecoproperties: processing

Polymer molding energy	*11.7	– 13.9	MJ/kg
Polymer molding CO_2	*0.93	– 1.1	kg/kg

Recycling

Recycle fraction in current supply	0.5	– 1	%

Typical uses. Electrical parts—sockets, switches, connectors, general industrial, water-lubricated bearings, relays, pump impellers, brake pistons, brake pads, microwave cookware, handles, bottles tops, coatings, adhesives, bearings, foams, and sandwich structures.

Natural rubber (NR)

The material. Natural rubber was known to the natives of Peru many centuries ago and is now one of Malaysia's main exports. It made the fortune of Giles Macintosh who, in 1825, devised the rubber-coated waterproof coat that still bears his name. Latex, the sap of the rubber tree, is cross-linked (vulcanized) by heating with sulfur; the amount of cross-linking determines the properties. It is the most widely used of all elastomers—more than 50% of all produced.

Composition
$(CH_2—C(CH_3)—CH—CH_2)_n$

General properties

Density	920	– 930	kg/m^3
Price	2.39	– 2.63	USD/kg

Mechanical properties

Young's modulus	0.0015	– 0.0025	GPa
Yield strength (elastic limit)	20	– 30	MPa
Tensile strength	22	– 32	MPa
Compressive strength	22	– 33	MPa
Elongation	500	– 800	%
Fatigue strength at 10^7 cycles	4.2	– 4.5	MPa
Fracture toughness	0.15	– 0.25	MPa.m$^{1/2}$

Thermal properties

Glass temperature	−78.2	– −63.2	°C
Maximum service temperature	68.9	– 107	°C
Thermal conductor or insulator?	Good insulator		
Thermal conductivity	0.1	– 0.14	W/m.K
Specific heat capacity	1800	– 2500	J/kg.K
Thermal expansion coefficient	150	– 450	µstrain/°C

Electrical properties

Electrical conductor or insulator?	Good insulator		
Electrical resistivity	1×10^{15}	– 1×10^{16}	µohm.cm
Dielectric constant	3	– 4.5	
Dissipation factor	7×10^{-4}	– 0.003	
Dielectric strength	16	– 23 10^6	V/m

Natural rubber is used in medical equipment, fashion items, tubing, and tires.

Ecoproperties: material

Annual world production	7.7×10^6	– 7.8×10^6	tonne/yr
Embodied energy, primary production	62	– 70	MJ/kg
CO_2 footprint, primary production	1.5	– 1.6	kg/kg
Water usage	*1500	– 2000	l/kg
Eco-indicator	340	– 380	millipoints/kg

Ecoproperties: processing

Polymer molding energy	*7.13	– 8.08	MJ/kg
Polymer molding CO_2	*0.57	– 0.646	kg/kg

Recycling

Recycle fraction in current supply	0.1	–	%

Typical uses. Gloves, car tires, seals, belts, antivibration mounts, electrical insulation, tubing, rubber lining pipes, and pumps.

Butyl rubber

The material. Butyl rubbers (BRs) are synthetics that resemble natural rubber in properties. They have good resistance to abrasion, tearing, and flexing, with exceptionally low gas permeability and useful properties up to 150°C. They have low dielectric constant and loss, making them attractive for electrical applications.

Composition
$(CH_2—C(CH_3)—CH—(CH_2)_2—C(CH_3)_2)_n$

General properties

Density	900	– 920	kg/m^3
Price	*3.93	– 4.32	USD/kg

Mechanical properties

Young's modulus	0.001	– 0.002	GPa
Yield strength (elastic limit)	2	– 3	MPa
Tensile strength	5	– 10	MPa
Compressive strength	2.2	– 3.3	MPa
Elongation	400	– 500	%
Fatigue strength at 10^7 cycles	*0.9	– 1.35	MPa
Fracture toughness	0.07	– 0.1	MPa.m$^{1/2}$

Thermal properties

Glass temperature	−73.2	– −63.2	°C
Maximum service temperature	96.9	– 117	°C
Thermal conductor or insulator?	Good insulator		
Thermal conductivity	0.08	– 0.1	W/m.K
Specific heat capacity	1.8e3	– 2.5e3	J/kg.K
Thermal expansion coefficient	120	– 300	µstrain/°C

Electrical properties

Electrical conductor or insulator?	Poor insulator		
Electrical resistivity	1×10^{15}	– 1×10^{16}	µohm.cm
Dielectric constant	*2.8	– 3.2	
Dissipation factor	0.001	– 0.01	
Dielectric strength	16	– 23 10^6	V/m

Butyl rubber is one of the most important materials for inner tubes.

Ecoproperties: material

Annual world production	1.03×10^7	– 1.06×10^7	tonne/yr
Reserves	*2.9×10^8	– 2.95×10^8	tonne
Embodied energy, primary production	95	– 120	MJ/kg
CO_2 footprint, primary production	3.6	– 4.2	kg/kg
Water usage	*63.8	– 191	l/kg

Ecoproperties: processing

Polymer molding energy	*7.24	– 8.08	MJ/kg
Polymer molding CO_2	*0.579	– 0.646	kg/kg

Recycling

Recycle fraction in current supply	2	– 4.1	%

Typical uses. Inner tubes, seals, belts, antivibration mounts, electrical insulation, tubing, brake pads, rubber lining pipes, and pumps.

EVA

The material. Ethylene-Vinyl-Acetate elastomers (EVA) are built around polyethylene. They are soft, flexible, and tough and retain these properties down to $-60°C$ and have good barrier properties as well as FDA approval for direct food contact. EVA can be processed by most normal thermoplastic processes: co-extrusion for films, blow molding, rotational molding, injection molding, and transfer molding.

Composition
$(CH_2)_n—(CH_2—CHR)_m$

General properties
Density	945	– 955	kg/m^3
Price	*2.1	– 2.31	USD/kg

Mechanical properties
Young's modulus	0.01	– 0.04	GPa
Yield strength (elastic limit)	12	– 18	MPa
Tensile strength	16	– 20	MPa
Compressive strength	13.2	– 19.8	MPa
Elongation	730	– 770	%
Fatigue strength at 10^7 cycles	*12	– 12.8	MPa
Fracture toughness	*0.5	– 0.7	MPa.m$^{1/2}$

Thermal properties
Glass temperature	* -73.2	– -23.2	°C
Maximum service temperature	46.9	– 51.9	°C
Thermal conductor or insulator?	Good insulator		
Thermal conductivity	0.3	– 0.4	W/m.K
Specific heat capacity	*2000	– 2200	J/kg.K
Thermal expansion coefficient	160	– 190	μstrain/°C

Electrical properties
Electrical conductor or insulator?	Good insulator		
Electrical resistivity	*3.1×10^{21}	– 1×10^{22}	μohm.cm
Dielectric constant	2.9	– 2.95	
Dissipation factor	0.005	– 0.022	
Dielectric strength	26.5	– 27 10^6	V/m

EVA is available in pastel or deep hues; it has good clarity and gloss. It has good barrier properties, little or no odor, and UV resistance.

Ecoproperties: material

Embodied energy, primary production	*86.7	– 95.8	MJ/kg
CO_2 footprint, primary production	*2.88	– 3.19	kg/kg
Water usage	*96.4	– 289	l/kg

Ecoproperties: processing

Polymer molding energy	*7.4	– 8.84	MJ/kg
Polymer molding CO_2	*0.592	– 0.707	kg/kg

Recycling

Embodied energy, recycling	*36.4	– 40.2	MJ/kg
CO_2 footprint, recycling	*1.21	– 1.34	kg/kg
Recycle fraction in current supply	6	– 10	%

Typical uses. Medical tubes, milk packaging, beer dispensing equipment, bags, shrink film, deep freeze bags, co-extruded and laminated film, closures, ice trays, gaskets, gloves, cable insulation, inflatable parts, running shoes.

Polychloroprene (Neoprene, CR)

The material. Polychloroprenes (Neoprene, CR), the materials of wetsuits, are the leading nontire synthetic rubbers. First synthesized in 1930, they are made by a condensation polymerization of the monomer 2-chloro–1,3 butadiene. The properties can by modified by copolymerization with sulfur, with other chloro-butadienes and by blending with other polymers to give a wide range of properties. Polychloroprenes are characterized by high chemical stability and resistance to water, oil, gasoline, and UV radiation.

Composition
$(CH_2—CCl—CH_2—CH_2)_n$

General properties

Density	1230	– 1250	kg/m^3
Price	*5.33	– 5.86	USD/kg

Mechanical properties

Young's modulus	7e-4	– 0.002	GPa
Yield strength (elastic limit)	3.4	– 24	MPa
Tensile strength	3.4	– 24	MPa
Compressive strength	3.72	– 28.8	MPa
Elongation	100	– 800	%
Fatigue strength at 10^7 cycles	*1.53	– 12	MPa
Fracture toughness	*0.1	– 0.3	$MPa.m^{1/2}$

Thermal properties

Glass temperature	−48.2	– −43.2	°C
Maximum service temperature	102	– 112	°C
Thermal conductor or insulator?	Good insulator		
Thermal conductivity	0.1	– 0.12	W/m.K
Specific heat capacity	*2000	– 2200	J/kg.K
Thermal expansion coefficient	575	– 610	μstrain/°C

Electrical properties

Electrical conductor or insulator?	Good insulator		
Electrical resistivity	1×10^{19}	– 1×10^{23}	μohm.cm
Dielectric constant	6.7	– 8	
Dissipation factor	*1×10^{-4}	– 0.001	
Dielectric strength	15.8	– 23.6 10^6	V/m

Neoprene gives wetsuits flexibility and stretch.

Ecoproperties: material

Embodied energy, primary production	*95.9	– 106	MJ/kg
CO_2 footprint, primary production	*3.4	– 3.9	kg/kg
Water usage	*126	– 378	l/kg

Ecoproperties: processing

Polymer molding energy	*7.77	– 8.57	MJ/kg
Polymer molding CO_2	*0.62	– 0.68	kg/kg

Recycling

Recycle fraction in current supply	*1	– 2	%

Typical uses. Brake seals, diaphragms, hoses and o-rings, tracked-vehicle pads, footwear, wetsuits.

12.4 Ceramics and glasses

Ceramics are materials of both the past and the future. They are the most durable of all materials—ceramic pots and ornaments survive from 5000 BC; Roman cement still bonds the walls of villas. It is ceramics' durability, particularly at high temperatures, that generates interest in them today. They are exceptionally hard (diamond, a ceramic, is the hardest of them all) and can tolerate higher temperatures than any metal. Ceramics are crystalline (or partly crystalline) inorganic compounds. They include high-performance technical ceramics such as alumina (used for electronic substrates, nozzles, and cutting tools), traditional, pottery-based ceramics (including brick and whiteware for baths, sinks, and toilets), and hydrated ceramics (cements and concretes) used for construction. All are hard and brittle and have generally high melting points and low thermal expansion coefficients, and most are good electrical insulators. When perfect they are exceedingly strong, but tiny flaws, hard to avoid, propagate as cracks when the material is loaded in tension or bending, drastically reducing the strength. The compressive strength, however, remains high (eight to 18 times the strength in tension). Impact resistance is low, and stresses due to thermal shock are not easily alleviated by plastic deformation, so large temperature gradients or thermal shock can cause failure.

Glass. Discovered by the Egyptians and perfected by the Romans, glass is one of the oldest manmade materials. For most of its long history it was a possession for the rich—as glass beads, ornaments, and vessels and as glaze on pottery. Its use in windows started in the 15th century, but it was not widespread until the 17th. Now, of course, it is so universal and cheap that, as bottles, we throw it away.

Glass is a mix of oxides, principally silica, SiO_2, that does not crystallize when cooled after melting. Pure glass is crystal-clear. Adding metal oxides produces a wide range of colors. Nickel gives a purple hue, cobalt a blue, chromium a green, uranium a green-yellow, iron a green–blue. The addition of iron gives a material that can absorb wavelengths in the infrared range so that heat radiation can be absorbed. Colorless, nonmetallic particles (fluorides or phosphates) are added from 5–15% to produce a translucent or an almost opaque white opalescence in glass and glass coatings. Photochromic glass changes color when exposed to UV. Filter glass protects from intense light and UV radiation; it is used in visors for welding.

Profiles for six ceramics follow, in the order in which they appear in Table 12.1.

Brick

The material. Brick is as old as Babylon (4000 BC) and as durable. It is the most ancient of all manmade building materials. The regularity and proportions of bricks make them easy to lay in a variety of patterns, and their durability makes them an ideal material for building construction. Clay, the raw material from which bricks are made, is available almost everywhere; finding the energy to fire them can be more of a problem. Pure clay is gray-white in color; the red color of most bricks comes from impurities of iron oxide.

Composition
Bricks are fired clays, fine particulate alumino-silicates that derive from the weathering of rocks.

General properties

Density	1600	– 2100	kg/m^3
Price	0.62	– 1.7	USD/kg

Mechanical properties

Young's modulus	15	– 30	GPa
Yield strength (elastic limit)	5	– 14	MPa
Tensile strength	5	– 14	MPa
Elongation	0		%
Hardness—Vickers	20	– 35	HV
Fatigue strength at 10^7 cycles	*6	– 9	MPa
Fracture toughness	1	– 2	MPa.m$^{1/2}$

Thermal properties

Melting point	927	– 1230	°C
Maximum service temperature	600	– 1000	°C
Thermal conductor or insulator?	Poor insulator		
Thermal conductivity	0.46	– 0.73	W/m.K
Specific heat capacity	750	– 850	J/kg.K
Thermal expansion coefficient	5	– 8	μstrain/°C

Electrical properties

Electrical conductor or insulator?	Good insulator		
Electrical resistivity	1×10^{14}	– 3×10^{16}	μohm.cm
Dielectric constant	7	– 10	
Dissipation factor	0.001	– 0.01	
Dielectric strength	9	– 15 10^6	V/m

The proportions and regularity of brick make it fast to assemble. Brick weathers well, and the texture and color make it visually attractive.

Ecoproperties: material

Annual world production	$^*5 \times 10^7$	– 5.1×10^7	tonne/yr
Embodied energy, primary production	2.2	– 3.5	MJ/kg
CO_2 footprint, primary production	0.2	– 0.23	kg/kg
Water usage	$^*2.8$	– 8.4	l/kg
Eco-indicator value	27	– 29	millipoints/kg

Ecoproperties: processing

Construction energy	$^*0.054$	– 0.066	MJ/kg
Construction CO_2	$^*0.009$	– 0.011	kg/kg

Recycling

Recycle fraction in current supply	15	– 20	%

Typical uses. Domestic and industrial building, walls, paths, and roads.

Stone

The material. Stone is the most durable of all building material. The Pyramids (before 3000 BC), the Parthenon (5th century BC), and the cathedrals of Europe (1000–1600 AD) testify to the resistance of stone to attack of every sort. It remained the principal material of construction for important buildings until the early 20th century; the railroads of the world, for example, could not have been built without stone for the viaducts and support structures. As the cost of stone increased and brick became cheaper, stone was increasingly used for the outer structure only; today it is largely used as a veneer on a concrete or breezeblock inner structure. Carefully selected samples of fully dense, defect-free stone can have very large compressive strengths—up to 1000MPa. But stone in bulk, as used in buildings, always contains defects. Then the average strength is much lower. The data given here is typical of bulk sandstone with a porosity of 5–30%. Bulk limestones are a little less strong, granites somewhat stronger.

Composition
There are many different compositions. The commonest are made up of calcium carbonate, silicates, and aluminates.

General properties

Density	2240	– 2650	kg/m^3
Price	0.3	– 1	USD/kg

Mechanical properties

Young's modulus	20	– 60	GPa
Yield strength (elastic limit)	2	– 25	MPa
Tensile strength	2	– 25	MPa
Elongation	0		%
Hardness—Vickers	*12	– 80	HV
Fatigue strength at 10^7 cycles	*2	– 18	MPa
Fracture toughness	0.7	– 1.4	MPa.m$^{1/2}$

Thermal properties

Melting point	1230	– 1430	°C
Maximum service temperature	350	– 900	°C
Thermal conductor or insulator?	Poor insulator		
Thermal conductivity	5.4	– 6	W/m.K
Specific heat capacity	840	– 920	J/kg.K
Thermal expansion coefficient	3.7	– 6.3	μstrain/°C

Stone, like wood, is one of man's oldest and most durable building materials.

Electrical properties

Electrical conductor or insulator?	Poor insulator	
Electrical resistivity	1×10^{10} – 1×10^{14}	μohm.cm
Dielectric constant	*6 – 9	
Dissipation factor	*0.001 – 0.01	
Dielectric strength	5 – 12 10^6	V/m

Ecoproperties: material

Embodied energy, primary production	4.9 – 6.4	MJ/kg
CO_2 footprint, primary production	*0.14 – 0.2	kg/kg
Water usage	*1.7 – 5.1	l/kg

Ecoproperties: processing

Construction energy	*0.036 – 0.044	MJ/kg
Construction CO_2	*0.0054 – 0.0066	kg/kg

Recycling

Recycle fraction in current supply	*1 – 2	%

Typical uses. Building and cladding, architecture, sculpture, optical benches for supports for high-performance or vibration-sensitive equipment such as microscopes.

Concrete

The material. Concrete is a composite, and a complex one. The matrix is cement; the reinforcement, a mixture of sand and gravel ("aggregate") occupying 60–80% of the volume. The aggregate increases the stiffness and strength and reduces the cost (aggregate is cheap). Concrete is strong in compression but cracks easily in tension. This is countered by adding steel reinforcement in the form of wire, mesh, or bars ("rebar"), often with surface contours to key it into the concrete; reinforced concrete can carry useful loads even when the concrete is cracked. Still higher performance is gained by using steel wire reinforcement that is pretensioned before the concrete sets. On relaxing the tension, the wires pull the concrete into compression.

Composition
6:1:2:4 Water:Portland cement:Fine aggregate:Coarse aggregate

General properties

Density	2300	– 2600	kg/m^3
Price	0.041	– 0.062	USD/kg

Mechanical properties

Young's modulus	15	– 25	GPa
Yield strength (elastic limit)	1	– 3	MPa
Tensile strength	1	– 1.5	MPa
Elongation	0		%
Hardness—Vickers	*5.7	– 6.3	HV
Fatigue strength at 10^7 cycles	*0.54	– 0.84	MPa
Fracture toughness	0.35	– 0.45	MPa.m$^{1/2}$

Thermal properties

Melting point	972	– 1230	°C
Maximum service temperature	480	– 510	°C
Thermal conductor or insulator?	Poor insulator		
Thermal conductivity	0.8	– 2.4	W/m.K
Specific heat capacity	835	– 1050	J/kg.K
Thermal expansion coefficient	6	– 13	μstrain/°C

Electrical properties

Electrical conductor or insulator?	Poor insulator		
Electrical resistivity	1.8×10^{12}	– 1.8×10^{13}	μohm.cm
Dielectric constant	*8	– 12	
Dissipation factor	*0.001	– 0.01	
Dielectric strength	0.8	– 1.8 10^6	V/m

Reinforced concrete enables large structures and complex shapes.

Ecoproperties: material

Annual world production	15×10^9	– 15.5×10^9	tonne/yr
Reserves	*500×10^9	– 510×10^9	tonne
Embodied energy, primary production	1	– 1.3	MJ/kg
CO_2 footprint, primary production	0.13	– 0.15	kg/kg
Water usage	*1.7	– 5.1	l/kg
Eco-indicator	3.6	– 4	millipoints/kg

1.15

≈ 1.2

.14

≈ 21

Ecoproperties: processing

Construction energy	*0.0182	– 0.022	MJ/kg
Construction CO_2	*0.0018	– 0.0022	kg/kg

.02

Recycling

Embodied energy, recycling	0.015	– 0.018	MJ/kg
CO_2 footprint, recycling	0.063	– 0.07	kg/kg
Recycle fraction in current supply	12.5	– 15	%

.017

Typical uses. General civil engineering construction and building.

Alumina

The material. Alumina (Al_2O_3) is to technical ceramics what mild steel is to metals—cheap, easy to process, the workhorse of the industry. It is the material of spark plugs, electrical insulators, and ceramic substrates for microcircuits. In single crystal form, it is sapphire, used for watch faces and cockpit windows of high-speed aircraft. More usually it is made by pressing and sintering powder, giving grades ranging from 80–99.9% alumina; the rest is porosity, glassy impurities, or deliberately added components. Pure aluminas are white; impurities make them pink or green. The maximum operating temperature increases with increasing alumina content. Alumina has a low cost and a useful and broad set of properties: electrical insulation, high mechanical strength, good abrasion, and temperature resistance up to 1650°C, excellent chemical stability, and moderately high thermal conductivity, but it has limited thermal shock and impact resistance. Chromium oxide is added to improve abrasion resistance; sodium silicate, to improve processability but with some loss of electrical resistance. Competing materials are magnesia, silica, and borosilicate glass.

Composition
Al_2O_3, often with some porosity and some glassy phase.

General properties
Density	3800	– 3980	kg/m^3
Price	*18.2	– 27.4	USD/kg

Mechanical properties
Young's modulus	343	– 390	GPa
Yield strength (elastic limit)	350	– 588	MPa
Tensile strength	350	– 588	MPa
Compressive strength	690	– 5.5e3	MPa
Elongation	0		%
Hardness-Vickers	1.2e3	– 2.06e3	HV
Fatigue strength at 10^7 cycles	*200	– 488	MPa
Fracture toughness	3.3	– 4.8	MPa.m$^{1/2}$

Thermal properties
Melting point	2000	– 2100	°C
Maximum service temperature	1080	– 1300	°C
Thermal conductor or insulator?	Good conductor		
Thermal conductivity	26	– 38.5	W/m.K
Specific heat capacity	790	– 820	J/kg.K
Thermal expansion coefficient	7	– 7.9	μstrain/°C

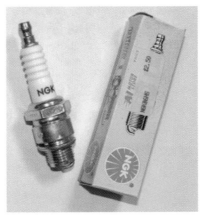

On the left: alumina components for wear resistance and for high temperature use (Kyocera Industrial Ceramics Corp.). On the right: an alumina spark plug insulator.

Electrical properties

Electrical conductor or insulator?	Good insulator	
Electrical resistivity	1×10^{20} – 1×10^{22}	μohm.cm
Dielectric constant	6.5 – 6.8	
Dissipation factor	1×10^{-4} – 4×10^{-4}	
Dielectric strength	10 – 20 10^{6}	V/m

Ecoproperties: material

Annual world production	1.19×10^{6} – 1.2×10^{6}	tonne/yr
Embodied energy, primary production	*49.5 – 54.7	MJ/kg
CO_2 footprint, primary production	*2.67 – 2.95	kg/kg
Water usage	*29.4 – 88.1	l/kg

104.2 (≈ 131)

2.82

≈ 4.95

Ecoproperties: processing

Ceramic powder forming energy	*25.3 – 27.8	MJ/kg
Ceramic powder forming CO_2	*2.02 – 2.23	kg/kg

27

2.13

Recycling

Recycle fraction in current supply	0.5 – 1	%

Typical uses. Electrical insulators and connector bodies; substrates; high temperature components; water faucet valves; mechanical seals; vacuum chambers and vessels; centrifuge linings; spur gears; fuse bodies; heating elements; plain bearings and other wear resistant components; cutting tools; substrates for microcircuits; spark plug insulators; tubes for sodium vapor lamps, thermal barrier coatings.

Soda-lime glass

The material. Soda-lime glass is the glass of windows, bottles, and light bulbs, used in vast quantities, the commonest of them all. The name suggests its composition: 13–17% NaO (the "soda"), 5–10% CaO (the "lime") and 70–75% SiO_2 (the "glass"). It has a low melting point, is easy to blow and mold, and it is cheap. It is optically clear unless impure, when it is typically green or brown. Windows today have to be flat and until 1950 that was not easy to do; now the float-glass process, solidifying glass on a bed of liquid tin, makes "plate" glass cheaply and quickly.

Composition
73% SiO_2/1% Al_2O_3/17% Na_2O/4% MgO/5% CaO

General properties

Density	2440	– 2490	kg/m^3
Price	*1.41	– 1.66	USD/kg

Mechanical properties

Young's modulus	68	– 72	GPa
Yield strength (elastic limit)	*30	– 35	MPa
Tensile strength	31	– 35	MPa
Elongation	0	–	%
Hardness—Vickers	439	– 484	HV
Fatigue strength at 10^7 cycles	*29.4	– 32.5	MPa
Fracture toughness	*0.55	– 0.7	$MPa.m^{1/2}$

Thermal properties

Maximum service temperature	443	– 673	K
Thermal conductor or insulator?	Poor insulator		
Thermal conductivity	*0.7	– 1.3	W/m.K
Specific heat capacity	*850	– 950	J/kg.K
Thermal expansion coefficient	9.1	– 9.5	μstrain/°C

Electrical properties

Electrical conductor or insulator?	Good insulator		
Electrical resistivity	7.94×10^{17}	– 7.94×10^{18}	μohm.cm
Dielectric constant	7	– 7.6	
Dissipation factor	0.007	– 0.01	
Dielectric strength	*12	– 14 10^6	V/m

Glass is used in both practical and decorative ways.

Eco properties: material

Annual world production	80×10^6	–	82×10^6	tonne/yr
Reserves	*10×10^9	–	11×10^9	tonne
Embodied energy, primary production	14	–	17	MJ/kg
CO_2 footprint, primary production	0.7	–	1	kg/kg
Water usage	*6.8	–	20.5	l/kg
Eco-indicator	48	–	53	millipoints/kg

Eco properties: processing

Glass molding energy	*7.82	–	9.46	MJ/kg
Glass molding CO_2	*0.62	–	0.75	kg/kg

Recycling

Embodied energy, recycling	6.16	–	7.48	MJ/kg
CO_2 footprint, recycling	0.31	–	0.44	kg/kg
Recycle fraction in current supply	22	–	26	%

Typical uses. Windows, bottles, containers, tubing, lamp bulbs, lenses and mirrors, bells, glazes on pottery and tiles.

Borosilicate glass (Pyrex)

The material. Borosilicate glass is soda lime glass with most of the lime replaced by borax, B_2O_3. It has a higher melting point than soda-lime glass and is harder to work, but it has a lower expansion coefficient and a high resistance to thermal shock, so it is used for glassware and laboratory equipment.

Composition
74% SiO_2/1% Al_2O_3/15% B_2O_3/4% Na_2O/6% PbO

General properties

Density	2200	– 2300	kg/m^3
Price	*4.15	– 6.22	USD/kg

Mechanical properties

Young's modulus	61	– 64	GPa
Yield strength (elastic limit)	*22	– 32	MPa
Tensile strength	22	– 32	MPa
Compressive strength	*264	– 384	MPa
Elongation	0		%
Hardness—Vickers	*83.7	– 92.5	HV
Fatigue strength at 10^7 cycles	*26.5	– 29.3	MPa
Fracture toughness	*0.5	– 0.7	MPa.m$^{1/2}$

Thermal properties

Glass temperature	450	– 602	°C
Maximum service temperature	230	– 460	°C
Thermal conductor or insulator?	Poor insulator		
Thermal conductivity	*1	– 1.3	W/m.K
Specific heat capacity	*760	– 800	J/kg.K
Thermal expansion coefficient	3.2	– 4	µstrain/°C

Electrical properties

Electrical conductor or insulator?	Good insulator		
Electrical resistivity	3.16×10^{21}	– 3.16×10^{22}	µohm.cm
Dielectric constant	4.65	– 6	
Dissipation factor	0.01	– 0.017	
Dielectric strength	*12	– 14 10^6	V/m

Borosilicate glass (Pyrex) is used for ovenware and chemical equipment.

Ecoproperties: material

Embodied energy, primary production	*23.8 – 26.3 MJ/kg
CO_2 footprint, primary production	*1.28 – 1.42 kg/kg
Water usage	*12.5 – 37.5 l/kg

Ecoproperties: processing

Glass molding energy	*7.42 – 8.98 MJ/kg
Glass molding CO_2	*0.59 – 0.72 kg/kg

Recycling

Embodied energy, recycling	*10.5 – 11.6 MJ/kg
CO_2 footprint, recycling	*0.56 – 0.62 kg/kg
Recycle fraction in current supply	*18 – 23 %

Typical uses. Ovenware, laboratory ware, piping, lenses and mirrors, sealed beam headlights, tungsten sealing, bells.

12.5 Hybrids: composites, foams, and natural materials

Composites are one of the great material developments of the 20th century. Those with the highest stiffness and strength are made with continuous fibers of glass, carbon, or Kevlar (an *aramid*) embedded in a thermosetting resin (polyester or epoxy). The fibers carry the mechanical loads, whereas the matrix material transmits loads to the fibers, provides ductility and toughness, and protects the fibers from damage from handling or the environment. It is the matrix material that limits the service temperature and processing conditions. Polyester-glass composites (GFRPs) are the cheapest, epoxy-carbon (CFRPs) and Kevlar-epoxy (KFRPs) the most expensive. A recent innovation is the use of thermoplastics as the matrix material, using a coweave of polypropylene and glass fibers that is thermoformed, melting the PP.

If continuous fiber CFRPs and GFRPs are the kings and queens of the composite world, the ordinary workers are polymers reinforced with chopped glass or carbon fibers (SMC and BMC) or with particulates (fillers) of silica sand, talc, or wood flour. They are used in far larger quantities, often in products so ordinary that most people would not guess that they were made of a composite: body panels of cars, household appliances, furniture, and fittings. It would, today, be hard to live without them.

So composites have remarkable potential. But the very thing that creates their properties—the hybridization of two very different materials—makes them near-impossible to recycle. In products with long lives, made in relatively small numbers (aircraft, for instance), this is not a concern; the fuel energy saved by the low weight of the composite far outweighs any penalty associated with the inability to recycle. But it is an obstacle to their use in high-volume, short-lived products (small cheap cars, for example).

Foams are made by variants of the process used to make bread. Mix an unpolymerized resin (the dough) with a hardener and a foaming agent (the yeast), wait for a bit, and the agent releases tiny gas bubbles that cause the mixture to rise in just the way bread does. There are other ways to make foams: violent stirring, like frothing egg white, or bubbling gas from below in the way you might make soap foam. All, suitably adapted, are used to make polymer foams. Those made from elastomers are soft and squashy, well adapted for cushions and packaging of delicate objects. Those made from thermoplastics or thermosets are rigid. They are used for more serious energy-absorbing and load-bearing applications: head protection in cycle helmets and cores for structural sandwich panels. And because they are mostly trapped gas, they are excellent thermal insulators.

Their ecocharacter, however, is mixed. The blowing agents used in the past—CFCs, chlorinated and fluoridated hydrocarbons—cause damage to the ozone layer; these have now been replaced. Some can be recycled, but only 1–10% of the foam that has to be collected, transported, and treated is real material (the rest is space), so you don't get much for your money.

Natural materials: wood, plywood, paper, and card. Wood has been used for construction since the earliest recorded time. The ancient Egyptians used it for furniture, sculpture, and coffins before 2500 BC. The Greeks at the peak of their empire (700 BC) and the Romans at the peak of theirs (around 0 AD) made elaborate buildings, bridges, boats, and chariots and weapons of wood and established the craft of furniture making that is still with us today. More diversity of use appeared in Medieval times, with the use of wood for large-scale building, and mechanisms such as carriages, pumps, windmills, even clocks, so that, right up to end of the 17th century, wood was the principal material of engineering. Since then cast iron, steel, and concrete have displaced it in some of its uses, but timber continues to be used on a massive scale, particularly in housing and small commercial buildings.

Plywood is laminated wood, the layers glued together such that the grains in successive layers are at right angles, giving stiffness and strength in both directions. The number of layers varies, but is always odd (3, 5, 7 …) to give symmetry about the core ply; if it is asymmetric it warps when wet or hot. Those with few plies (3, 5) are significantly stronger and stiffer in the direction of the outermost layers; with increasing number of plies the properties become more uniform. High-quality plywood is bonded with synthetic resin. The data listed here describe the in-plane properties of a typical five-ply.

Papyrus, the forerunner of paper, was made from the flower stem of the reed, native to Egypt; it has been known and used for over 5000 years. Paper, by contrast, is a Chinese invention (105 AD). It is made from pulped cellulose fibers derived from wood, cotton, or flax. Paper making uses caustic soda (NaOH) and vast quantities of water—a bad combination if released back into the environment. Modern paper-making plants now release water that (they claim) is as clean as when it entered.

Profiles for 12 hybrid materials follow, in the order in which they appear in Table 12.1.

CFRP (Isotropic)

The material. Carbon fiber reinforced composites (CFRPs) offer greater stiffness and strength than any other type, but they are considerably more expensive than GFRP (see record). Continuous fibers in a polyester or epoxy matrix give the highest performance. The fibers carry the mechanical loads, whereas the matrix material transmits loads to the fibers and provides ductility and toughness as well as protecting the fibers from damage caused by handling or the environment. It is the matrix material that limits the service temperature and processing conditions.

Composition
Epoxy + continuous HS carbon fiber reinforcement (0, + − 45, 90), quasi-isotropic layup.

General properties
Density	1500	– 1600	kg/m^3
Price	*40.0	– 44.0	USD/kg

Mechanical properties
Young's modulus	69	– 150	GPa
Yield strength (elastic limit)	550	– 1050	MPa
Tensile strength	550	– 1050	MPa
Elongation	*0.32	– 0.35	%
Hardness—Vickers	*10.8	– 21.5	HV
Fatigue strength at 10^7 cycles	*150	– 300	MPa
Fracture toughness	*6.12	– 20	MPa.m$^{1/2}$

Thermal properties
Maximum service temperature	*140	– 220	°C
Thermal conductor or insulator?	Poor insulator		
Thermal conductivity	*1.28	– 2.6	W/m.K
Specific heat capacity	*902	– 1037	J/kg.K
Thermal expansion coefficient	*1	– 4	μstrain/°C

Electrical properties
Electrical conductor or insulator?	Poor conductor		
Electrical resistivity	*1.65×10^5	– 9.46×10^5	μohm.cm

A CFRP bike frame. (Courtesy TREK.)

Ecoproperties: material

Annual world production	2.8×10^4	2.85×10^4	tonne/yr
Embodied energy, primary production	*259	– 286	MJ/kg
CO_2 footprint, primary production	*16.1	– 18.5	kg/kg
Water usage	*360	– 1367	l/kg

Ecoproperties: processing

Simple composite molding energy	*11.3	– 12.4	MJ/kg
Simple composite molding CO_2	*0.90	– 0.99	kg/kg
Advanced composite molding energy	*18.4	– 20.3	MJ/kg
Advanced composite molding CO_2	*1.48	– 1.63	kg/kg

Recycling

Recycle fraction in current supply	0.5	– 1	%

Typical uses. Lightweight structural members in aerospace, ground transport, and sports equipment such as bikes, golf clubs, oars, boats, and racquets; springs; pressure vessels.

GFRP (Isotropic)

The material. Composites are one of the great material developments of the 20th century. Those with the highest stiffness and strength are made of continuous fibers (glass, carbon, or Kevlar, an aramid) embedded in a thermosetting resin (polyester or epoxy). The fibers carry the mechanical loads, whereas the matrix material transmits loads to the fibers and provides ductility and toughness as well as protecting the fibers from damage caused by handling or the environment. It is the matrix material that limits the service temperature and processing conditions. Polyester-glass composites (GFRPs) are the cheapest and by far the most widely used. A recent innovation is the use of thermoplastics at the matrix material, either in the form of a coweave of cheap polypropylene and glass fibers that is thermoformed, melting the PP, or as expensive high-temperature thermoplastic resins such as PEEK that allow composites with higher temperature and impact resistance. High-performance GFRP uses continuous fibers. Those with chopped glass fibers are cheaper and are used in far larger quantities. GFRP products range from tiny electronic circuit boards to large boat hulls, body and interior panels of cars, household appliances, furniture, and fittings.

Composition
Epoxy + continuous E-glass fiber reinforcement (0, + − 45, 90), quasi-isotropic layup.

General properties

Density	1750 – 1970	kg/m^3
Price	*19.44 – 21.39	USD/kg

Mechanical properties

Young's modulus	*15 – 28	GPa
Yield strength (elastic limit)	*110 – 192	MPa
Tensile strength	*138 – 241	MPa
Elongation	*0.85 – 0.95	%
Hardness—Vickers	*10.8 – 21.5	HV
Fatigue strength at 10^7 cycles	*55 – 96	MPa
Fracture toughness	*7 – 23	MPa.m$^{1/2}$

Thermal properties

Maximum service temperature	*413 – 493	°C
Thermal conductor or insulator?	Poor insulator	
Thermal conductivity	*0.4 – 0.55	W/m.K
Specific heat capacity	*1000 – 1200	J/kg.K
Thermal expansion coefficient	*8.6 – 32.9	μstrain/°C

GFRP body shell by MAS Design, Windsor, UK.

Electrical properties

Electrical conductor or insulator?	Good insulator			
Electrical resistivity	*2.4 × 10^{21}	–	1.91 × 10^{22}	μohm.cm
Dielectric constant	4.86	–	5.17	
Dissipation factor	0.004	–	0.009	
Dielectric strength	11.8	–	19.7 10^{6}	V/m

Electrical conductor or insulator?   Good insulator
Electrical resistivity $^{*}2.4 \times 10^{21}$ – 1.91×10^{22} μohm.cm
Dielectric constant   4.86 – 5.17
Dissipation factor   0.004 – 0.009
Dielectric strength   11.8 – 19.7 10^{6} V/m

Ecoproperties

Embodied energy, primary production $^{*}107$ – 118   MJ/kg
CO_2 footprint, primary production $^{*}7.47$ – 8.26   kg/kg
Water usage $^{*}105$ – 309   l/kg

Material processing: energy

Simple composite molding energy $^{*}11.3$ – 12.4   MJ/kg
Simple composite molding CO_2 $^{*}0.90$ – 0.99   kg/kg
Advanced composite molding energy $^{*}18.4$ – 20.3   MJ/kg
Advanced composite molding CO_2 $^{*}1.48$ – 1.63   kg/kg

Material recycling: energy, CO_2 and recycle fraction

Recycle fraction in current supply   0.5 – 1   %

Typical uses. Sports equipment such as skis, racquets, skate boards and golf club shafts; ship and boat hulls; body shells; automobile components; cladding and fittings in construction; chemical plant.

Sheet molding compound (SMC)

The material. Layup and filament winding methods of shaping composites are far too slow and labor intensive to compete with steel pressings for car body panels and other enclosures. Sheet molding compounds (SMCs) overcome this by allowing molding in a single operation between heated dies. To make SMC, polyester resin containing thickening agents and cheap particulates such as calcium carbonate or silica dust is mixed with chopped fibers—usually glass—to form a sheet. The fibers lie more or less parallel to the plane of the sheet but are randomly oriented in-plane, with a volume fraction between 15% and 40%. This makes a "pre-preg" with leather- or doughlike consistency. When an SMC sheet is pressed between hot dies, it polymerizes, giving a strong, stiff sheet molding.

Composition
$(OOC—C_6H_4—COO—C_6H_{10})_n$ + $CaCO_3$ or SiO_2 filler + 15 to 40% chopped-glass strand.

General properties

Density	1800	– 2000	kg/m^3
Price	*5.81	– 6.39	USD/kg

Mechanical properties

Young's modulus	9	– 14	GPa
Yield strength (elastic limit)	50	– 90	MPa
Tensile strength	60	– 100	MPa
Compressive strength	240	– 310	MPa
Elongation	2.5	– 3.2	%
Hardness—Vickers	*15	– 25	HV
Fatigue strength at 10^7 cycles	*20	– 36	MPa
Fracture toughness	*5	– 13	MPa.m$^{1/2}$

Thermal properties

Glass temperature	147	– 197	°C
Maximum service temperature	180	– 220	°C
Thermal conductor or insulator?	Good insulator		
Thermal conductivity	0.27	– 0.5	W/m.K
Specific heat capacity	1050	– 1090	J/kg.K
Thermal expansion coefficient	18	– 33	μstrain/°C

SMC cycle shed. (Image courtesy of the ACT Program, McMaster University.)

Electrical properties

Electrical conductor or insulator?	Good insulator			
Electrical resistivity	1×10^{18}	–	1×10^{19}	μohm.cm
Dielectric constant	4.25	–	5.1	
Dissipation factor	0.003	–	0.0095	
Dielectric strength	10	–	18 10^6	V/m

Ecoproperties: material

Embodied energy, primary production	*110	–	120	MJ/kg
CO_2 footprint, primary production	*7.7	–	8.5	kg/kg
Water usage	*89	–	280	l/kg

Eco properties: processing

Simple composite molding energy	*11.3	–	12.4	MJ/kg
Simple composite molding CO_2	*0.90	–	0.99	kg/kg
Advanced composite molding energy	*18.4	–	20.3	MJ/kg
Advanced composite molding CO_2	*1.48	–	1.63	kg/kg

Recycling

Recycle fraction in current supply	0.5	–	1	%

Typical uses. Sheet pressings of all types, competing with steel and aluminum sheet. Car body panels; enclosures; luggage and packing cases.

Bulk molding compound (BMC)

The material. Layup and filament winding methods of shaping composites are far too slow and labor-intensive to compete with steel pressings for car body panels and other enclosures. Sheet molding compounds (SMCs) and bulk (or dough) molding compounds (BMCs or DMCs) overcome this by allowing molding in a single operation between heated dies. To make BMC, polyester resin containing thickening agents and cheap particulates such as calcium carbonate or silica dust is mixed with chopped fibers—usually glass—to form a sheet. The fibers lie more or less parallel to the plane of the sheet but are randomly oriented in three dimensions, with a volume fraction between 10% and 25%. This makes a "pre-preg" with leather- or doughlike consistency. BMC is molded in closed, heated dies to make more complex shapes: door handles, shaped levers, parts for washing machines, and the like.

Composition
$(OOC—C_6H_4—COO—C_6H_{10})_n$ + $CaCO_3$ or SiO_2 filler + 15 to 40% chopped-glass strand.

General properties

Density	1800	– 2100	kg/m^3
Price	*4.75	– 5.22	USD/kg

Mechanical properties

Young's modulus	12	– 14	GPa
Yield strength (elastic limit)	25	– 55	MPa
Tensile strength	34	– 70	MPa
Compressive strength	140	– 180	MPa
Elongation	1.4	– 1.9	%
Hardness—Vickers	*7	– 16	HV
Fatigue strength at 10^7 cycles	*12	– 27	MPa
Fracture toughness	*3	– 6	MPa.m$^{1/2}$

Thermal properties

Glass temperature	147	– 197	°C
Maximum service temperature	140	– 210	°C
Thermal conductor or insulator?	Good insulator		
Thermal conductivity	0.27	– 0.5	W/m.K
Specific heat capacity	*1110	– 1160	J/kg.K
Thermal expansion coefficient	24	– 34	μstrain/°C

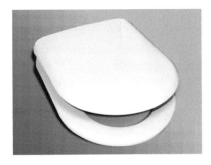

A BMC (or DMC) molding. BMC is used for door handles, casings for electrical and gas, and most small moldings in cars.

Electrical properties

Electrical conductor or insulator?	Good insulator	
Electrical resistivity	1×10^{18} – 1×10^{19}	μohm.cm
Dielectric constant	4.2 – 5	
Dissipation factor	0.002 – 0.008	
Dielectric strength	10 – 18 10^6	V/m

Ecoproperties: material

Embodied energy, primary production	109 – 121	MJ/kg
CO_2 footprint, primary production	7.5 – 8.5	kg/kg
Water usage	*89 – 280	l/kg

Ecoproperties: processing

Simple composite molding energy	*11.3 – 12.4	MJ/kg
Simple composite molding CO_2	*0.90 – 0.99	kg/kg
Advanced composite molding energy	*18.4 – 20.3	MJ/kg
Advanced composite molding CO_2	*1.48 – 1.63	kg/kg

Recycling

Recycle fraction in current supply	0.5 – 1	%

Typical uses. Car battery cases; door handles and window winders; washing machine parts such as lids; automotive vents, distributor caps and other small moldings; casings for telephones, gas, and electricity meters.

Rigid polymer foam

The material. Polymer foams are made by the controlled expansion and solidification of a liquid or melted through a blowing agent; physical, chemical, or mechanical blowing agents are possible. The resulting cellular material has a lower density, stiffness, and strength than the parent material, by an amount that depends on its relative density—the volume fraction of solid in the foam. Rigid foams are made from polystyrene, phenolic, polyethylene, polypropylene, or derivatives of polymethylmethacrylate. They are light and stiff and have mechanical properties that make them attractive for energy management and packaging and for lightweight structural use. Open-cell foams can be used as filters, closed-cell foams as flotation. Self-skinning foams, called *structural* or *syntactic*, have a dense surface skin made by foaming in a cold mold. Rigid polymer foams are widely used as cores of sandwich panels.

Composition
Hydrocarbon.

General properties
Density	78	– 165	kg/m^3
Price	*12	– 24	USD/kg

Mechanical properties
Young's modulus	0.08	– 0.2	GPa
Yield strength (elastic limit)	0.4	– 3.5	MPa
Tensile strength	0.65	– 5.1	MPa
Compressive strength	0.95	– 3.5	MPa
Elongation	2	– 5	%
Hardness—Vickers	0.095	– 0.35	HV
Fatigue strength at 10^7 cycles	*0.455	– 2.8	MPa
Fracture toughness	*0.0066	– 0.048	MPa.m$^{1/2}$

Thermal properties
Glass temperature	67	– 157	°C
Maximum service temperature	67	– 157	°C
Thermal conductor or insulator?	Good insulator		
Thermal conductivity	0.027	– 0.038	W/m.K
Specific heat capacity	1120	– 1910	J/kg.K
Thermal expansion coefficient	20	– 70	μstrain/°C

Rigid polymer foam is used as the core of the GFRP sandwich shell for ultra-lightweight designs such as this glider.

Electrical properties

Electrical conductor or insulator?	Good insulator
Electrical resistivity	1×10^{17} – 1×10^{21} µohm.cm
Dielectric constant	1.1 – 1.19
Dissipation factor	8×10^{-4} – 0.008
Dielectric strength	5.6 – 6.7 10^6 V/m

Ecoproperties: material

Embodied energy, primary production	*105 – 110 MJ/kg
CO_2 footprint, primary production	*3.5 – 4.2 kg/kg
Water usage	*299 – 865 l/kg
Eco-indicator	400 – 440 millipoints/kg

Ecoproperties: processing

Polymer molding energy	* 9.75 – 11.8 MJ/kg
Polymer molding CO_2	* 0.78 – 0.94 kg/kg
Polymer extrusion energy	* 3.77 – 4.56 MJ/kg
Polymer extrusion CO_2	* 0.3 – 0.36 kg/kg

Recycling

Recycle fraction in current supply	0.1 %

Typical uses. Thermal insulation, cores for sandwich structures, panels, partitions, refrigeration, energy absorption, packaging, buoyancy, floatation.

Flexible polymer foam

The material. Polymer foams are made by the controlled expansion and solidification of a liquid or melt through a blowing agent; physical, chemical, or mechanical blowing agents are possible. The resulting cellular material has a lower density, stiffness, and strength than the parent material, by an amount that depends on its relative density—the volume fraction of solid in the foam. Flexible foams can be soft and compliant, the material of cushions, mattresses, and padded clothing. Most are made from polyurethane, although latex (natural rubber) and most other elastomers can be foamed.

Composition
Hydrocarbon.

General properties

Density	70	– 115	kg/m^3
Price	*3.11	– 3.32	USD/kg

Mechanical properties

Young's modulus	0.004	– 0.012	GPa
Yield strength (elastic limit)	0.048	– 0.7	MPa
Tensile strength	0.43	– 2.95	MPa
Compressive strength	0.048	– 0.7	MPa
Elongation	9	– 115	%
Hardness—Vickers	0.0048	– 0.07	HV
Fatigue strength at 10^7 cycles	*0.34	– 2.5	MPa
Fracture toughness	*0.03	– 0.09	MPa.m$^{1/2}$

Thermal properties

Melting point	112	– 177	°C
Glass temperature	−113	– −13.2	°C
Maximum service temperature	82.9	– 112	°C
Thermal conductor or insulator?	Good insulator		
Thermal conductivity	0.041	– 0.078	W/m.K
Specific heat capacity	1750	– 2260	J/kg.K
Thermal expansion coefficient	115	– 220	µstrain/°C

Electrical properties

Electrical conductor or insulator?	Good insulator		
Electrical resistivity	1×10^{20}	– 1×10^{21}	µohm.cm
Dielectric constant	1.2	– 1.3	
Dissipation factor	5×10^{-4}	– 0.003	
Dielectric strength	4	– 6 10^6	V/m

Flexible latex foams are used for cushions, mattresses, and packaging.

Ecoproperties: material

Embodied energy, primary production	*104	– 115	MJ/kg
CO_2 footprint, primary production	*4	– 4.8	kg/kg
Water usage	*181	– 544	l/kg
Eco-indicator	460	– 500	millipoints/kg

Ecoproperties: processing

Polymer molding energy	*6.92	– 8.38	MJ/kg
Polymer molding CO_2	*0.55	– 0.67	kg/kg
Polymer extrusion energy	*2.71	– 3.28	MJ/kg
Polymer extrusion CO_2	*0.21	– 0.26	kg/kg

Recycling

Recycle fraction in current supply	0.1	%

Typical uses. Packaging, buoyancy, cushioning, sleeping mats, soft furnishings, artificial skin, sponges, carriers for inks and dyes.

Paper and cardboard

The material. Papyrus, the forerunner of paper, was made from the flower stem of the reed, native to Egypt; it has been known and used for over 5000 years. Paper, by contrast, is a Chinese invention (105 AD). It is made from pulped cellulose fibers derived from wood, cotton, or flax. There are many types of paper and paperboard: tissue paper, newsprint, Kraft paper for packaging, office paper, fine glazed writing paper, cardboard—and a correspondingly wide range of properties. The following data spans the range of newsprint and Kraft paper.

Composition
Cellulose fibers, usually with filler and colorant.

General properties
Density	480	– 860	kg/m^3
Price	2.07	– 12.4	USD/kg

Mechanical properties
Young's modulus	3	– 8.9	GPa
Yield strength (elastic limit)	15	– 34	MPa
Tensile strength	23	– 51	MPa
Compressive strength	41	– 55	MPa
Elongation	0.75	– 2	%
Hardness—Vickers	*4	– 9	HV
Fatigue strength at 10^7 cycles	*13	– 24	MPa
Fracture toughness	*6	– 10	MPa.m$^{1/2}$

Thermal properties
Glass temperature	47	– 67	°C
Maximum service temperature	77	– 130	°C
Thermal conductor or insulator?	Good conductor		
Thermal conductivity	0.06	– 0.17	W/m.K
Specific heat capacity	1340	– 1400	J/kg.K
Thermal expansion coefficient	5	– 20	μstrain/°C

Electrical properties
Electrical conductor or insulator?	Good insulator		
Electrical resistivity	1×10^{13}	– 1×10^{14}	μohm.cm
Dielectric constant	2.5	– 6	
Dissipation factor	0.015	– 0.04	
Dielectric strength	0.2	– 0.3 10^6	V/m

Cardboard, ready for recycling.

Ecoproperties: material

Annual world production	3.6×10^8	$- 3.7 \times 10^8$	tonne/yr
Embodied energy, primary production	24.2	– 32	MJ/kg
CO_2 footprint, primary production	1.23	– 1.55	kg/kg
Water usage	*1100	– 1200	l/kg
Eco-indicator	91	– 100	millipoints/kg

Ecoproperties: processing

Construction energy	*0.475	– 0.525	MJ/kg
Construction CO_2	*0.023	– 0.026	kg/kg

Recycling

Embodied energy, recycling	18	– 20	MJ/kg
CO_2 footprint, recycling	*0.72	– 0.78	kg/kg
Recycle fraction in current supply	70	– 74	%

Typical uses. Packaging, filtering, writing; printing; currency; electrical and thermal insulation; gaskets.

Plywood

The material. Plywood is laminated wood, the layers glued together such that the grain in successive layers are at right angles, giving stiffness and strength in both directions. The number of layers varies but is always odd (3, 5, 7 ...) to give symmetry about the core ply; if it is asymmetric it warps when wet or hot. Those with few plies (3, 5) are significantly stronger and stiffer in the direction of the outermost layers; with increasing number of plies the properties become more uniform. High-quality plywood is bonded with synthetic resin. The following data describes the in-plane properties of a typical five-ply.

Composition
Cellulose/hemicellulose/lignin/12%H_2O/adhesive.

General properties

Density	700	– 800	kg/m^3
Price	1.04	– 2.07	USD/kg

Mechanical properties

Young's modulus	6.9	– 13	GPa
Yield strength (elastic limit)	*9	– 30	MPa
Tensile strength	10	– 44	MPa
Compressive strength	8	– 25	MPa
Elongation	2.4	– 3	%
Hardness—Vickers	3	– 9	HV
Fatigue strength at 10^7 cycles	*7	– 16	MPa
Fracture toughness	*1	– 1.8	MPa.m$^{1/2}$

Thermal properties

Glass temperature	120	– 140	°C
Maximum service temperature	*100	– 130	°C
Thermal conductor or insulator?	Good insulator		
Thermal conductivity	0.3	– 0.5	W/m.K
Specific heat capacity	1660	– 1710	J/kg.K
Thermal expansion coefficient	6	– 8	μstrain/°C

Electrical properties

Electrical conductor or insulator?	Poor insulator		
Electrical resistivity	6×10^{13}	– 2×10^{14}	μohm.cm
Dielectric constant	6	– 8	
Dissipation factor	*0.05	– 0.09	
Dielectric strength	0.4	0.6 10^6	V/m

Plywood dominates the market for both wood and steel stud construction. It is widely used, too, for furniture and fittings, boat building, and packaging.

Ecoproperties: material

Embodied energy, primary production	13	– 17	MJ/kg
CO_2 footprint, primary production	0.73	– 0.77	kg/kg
Water usage	*500	– 1000	l/kg
Eco-indicator	37	– 41	millipoints/kg

Ecoproperties: processing

Construction energy	*0.455	– 0.55	MJ/kg
Construction CO_2	*0.022	– 0.027	kg/kg

Recycling

Recycle fraction in current supply	1	– 2	%

Typical uses. Furniture, building and construction, marine and boat building, packaging, transport and vehicles, musical instruments, aircraft, modeling.

Softwood: pine, along grain

The material. Softwoods come from coniferous, mostly evergreen, trees such as spruce, pine, fir, and redwood. Wood must be seasoned before it is used. Seasoning is the process of drying to remove some of the natural moisture to make it dimensionally stable. In air seasoning the wood is dried naturally in covered but open-sided structures. In kiln drying the wood is artificially dried in an oven. Wood has been used for construction and to make products since the earliest times. Timber continues to be used on a massive scale, particularly in housing and commercial buildings.

Composition
Cellulose/hemicellulose/lignin/12%H_2O.

General properties

Density	440	– 600	kg/m^3
Price	1.04	– 2.07	USD/kg

Mechanical properties

Young's modulus	8.4	– 10.3	GPa
Yield strength (elastic limit)	*35	– 45	MPa
Tensile strength	*60	– 100	MPa
Compressive strength	*35	– 43	MPa
Elongation	*1.99	– 2.43	%
Hardness—Vickers	*3	– 4	HV
Fatigue strength at 10^7 cycles	*19	– 23	MPa
Fracture toughness	*3.4	– 4.1	MPa.m$^{1/2}$

Thermal properties

Glass temperature	77	– 102	°C
Maximum service temperature	120	– 140	°C
Thermal conductor or insulator?	Good insulator		
Thermal conductivity	*0.22	– 0.3	W/m.K
Specific heat capacity	1660	– 1710	J/kg.K
Thermal expansion coefficient	*2.5	– 9	μstrain/°C

Electrical properties

Electrical conductor or insulator?	Poor insulator		
Electrical resistivity	*6×10^{13}	– 2×10^{14}	μohm.cm
Dielectric constant	*5	– 6.2	
Dissipation factor	*0.05	– 0.1	
Dielectric strength	*0.4	0.6 10^6	V/m

 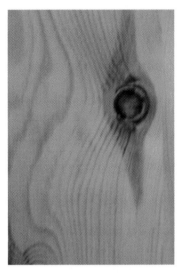

Wood remains one of the world's major structural materials as well finding application in more delicate objects such as furniture and musical instruments.

Ecoproperties: material

Annual world production	9.6×10^8	$- 9.7 \times 10^8$	tonne/yr
Embodied energy, primary production	7	– 7.8	MJ/kg
CO_2 footprint, primary production	0.4	– 0.46	kg/kg
Water usage	*500	– 750	l/kg
Eco-indicator	6.3	– 6.9	millipoints/kg

Ecoproperties: processing

Construction energy	*0.455	– 0.55	MJ/kg
Construction CO_2	*0.022	– 0.027	kg/kg

Recycling

Recycle fraction in current supply	8	– 10	%

Typical uses. Flooring; furniture; containers; cooperage; sleepers (when treated); building construction; boxes; crates and palettes; planing-mill products; subflooring; sheathing; and as the feedstock for plywood, particle-board, and hardboard.

Softwood: pine, across grain

The material. Softwoods come from coniferous, mostly evergreen, trees such as spruce, pine, fir, and redwood. Wood must be seasoned before it is used. Seasoning is the process of drying to remove some of the natural moisture to make it dimensionally stable. In air seasoning the wood is dried naturally in covered but open-sided structures. In kiln drying the wood is artificially dried in an oven. Wood has been used for construction and to make products since the earliest times. Timber continues to be used on a massive scale, particularly in housing and commercial buildings.

Composition
Cellulose/hemicellulose/lignin/12%H_2O.

General properties

Density	440	– 600	kg/m^3
Price	1.04	– 2.07	USD/kg

Mechanical properties

Young's modulus	0.6	– 0.9	GPa
Yield strength (elastic limit)	*1.7	– 2.6	MPa
Tensile strength	3.2	– 3.9	MPa
Compressive strength	*3	– 9	MPa
Elongation	1	– 1.5	%
Hardness—Vickers	2.6	– 3.2	HV
Fatigue strength at 10^7 cycles	*0.96	– 1.2	MPa
Fracture toughness	*0.4	– 0.5	$MPa.m^{1/2}$

Thermal properties

Glass temperature	77	– 102	°C
Maximum service temperature	120	– 140	°C
Thermal conductor or insulator?	Good insulator		
Thermal conductivity	0.08	– 0.14	W/m.K
Specific heat capacity	1660	– 1710	J/kg.K
Thermal expansion coefficient	*26	– 36	μstrain/°C

Electrical properties

Electrical conductor or insulator?	Poor insulator		
Electrical resistivity	*2.1×10^{14}	– 7×10^{14}	μohm.cm
Dielectric constant	*5	– 6.2	
Dissipation factor	*0.03	– 0.07	
Dielectric strength	1	– $2 \; 10^6$	V/m

 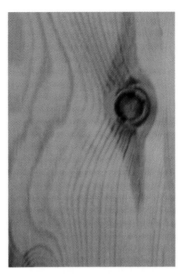

Wood remains one of the world's major structural materials as well finding application in more delicate objects such as furniture and musical instruments.

Ecoproperties: material

Annual world production	9.6×10^8	– 9.7×10^8	tonne/yr
Embodied energy, primary production	7	– 7.8	MJ/kg
CO_2 footprint, primary production	0.4	– 0.46	kg/kg
Water usage	*500	– 750	l/kg
Eco-indicator	6.3	– 6.9	millipoints/kg

Ecoproperties: processing

Construction energy	*0.455	– 0.55	MJ/kg
Construction CO_2	*0.022	– 0.027	kg/kg

Recycling

Recycle fraction in current supply	8	– 10	%

Typical uses. Flooring; furniture; containers; cooperage; sleepers (when treated); building construction; boxes; crates and palettes; planing-mill products; subflooring; sheathing and as the feedstock for plywood, particleboard, and hardboard.

Hardwood: oak, along grain

The material. Hardwoods come from broad-leaved, deciduous trees such as oak, ash, elm, sycamore, and mahogany. Although most hardwoods are harder than softwoods, there are exceptions: balsa, for instance, is a hardwood. Wood must be seasoned before it is used. Seasoning is the process of drying the natural moisture out of the raw timber to make it dimensionally stable, allowing its use without shrinking or warping. In air seasoning the wood is dried naturally in covered but open-sided structures. In kiln drying the wood is artificially dried in an oven or kiln.

Composition
Cellulose/hemicellulose/lignin/12%H_2O.

General properties

Density	850	– 1030	kg/m^3
Price	3.11	– 4.15	USD/kg

Mechanical properties

Young's modulus	20.6	– 25.2	GPa
Yield strength (elastic limit)	43	– 52	MPa
Tensile strength	132	– 162	MPa
Compressive strength	68	– 83	MPa
Elongation	*1.7	– 2.1	%
Hardness—Vickers	*13	– 15.8	HV
Fatigue strength at 10^7 cycles	*42	– 52	MPa
Fracture toughness	*9	– 10	MPa.m$^{1/2}$

Thermal properties

Glass temperature	77	– 102	°C
Maximum service temperature	120	– 140	°C
Thermal conductor or insulator?	Good insulator		
Thermal conductivity	*0.41	– 0.5	W/m.K
Specific heat capacity	1660	– 1710	J/kg.K
Thermal expansion coefficient	*2.5	– 9	µstrain/°C

Electrical properties

Electrical conductor or insulator?	Poor insulator		
Electrical resistivity	*6 × 10^{13}	– 2 × 10^{14}	µohm.cm
Dielectric constant	*5	– 6	
Dissipation factor	*0.1	– 0.15	
Dielectric strength	*0.4	– 0.6 10^6	V/m

Wood remains one of the world's major structural materials as well finding application in more delicate objects such as furniture.

Ecoproperties: material

Embodied energy, primary production	7.4	– 8.2	MJ/kg
CO_2 footprint, primary production	0.45	– 0.49	kg/kg
Water usage	*730	– 1.5e3	l/kg
Eco-indicator	6.3	– 6.9	millipoints/kg

Ecoproperties: processing

Construction energy	*0.909	– 1.1	MJ/kg
Construction CO_2	*0.045	– 0.05	kg/kg

Recycling

Recycle fraction in current supply	8	– 10	%

Typical uses. Flooring; stairways, furniture; handles; veneer; sculpture; woodenware; sashes; doors; general millwork; framing—but these are just a few. Almost every load-bearing and decorative object has, at one time or another, been made from wood.

Hardwood: oak, across grain

The material. Hardwoods come from broad-leaved, deciduous trees such as oak, ash, elm, sycamore, and mahogany. Although most hardwoods are harder than softwoods, there are exceptions: balsa, for instance, is a hardwood. Wood must be seasoned before it is used. Seasoning is the process of drying the natural moisture out of the raw timber to make it dimensionally stable, allowing its use without shrinking or warping. In air seasoning the wood is dried naturally in covered but open-sided structures. In kiln drying the wood is artificially dried in an oven or kiln.

Composition
Cellulose/hemicellulose/lignin/12%H_2O.

General properties

Density	850	– 1.13e3	kg/m^3
Price	3.11	– 4.15	USD/kg

Mechanical properties

Young's modulus	4.5	– 5.8	GPa
Yield strength (elastic limit)	*4	– 5.9	MPa
Tensile strength	7.1	– 8.7	MPa
Compressive strength	*12.7	– 15.6	MPa
Elongation	1	– 1.5	%
Hardness—Vickers	10	– 12	HV
Fatigue strength at 10^7 cycles	*2.1	– 2.6	MPa
Fracture toughness	*0.8	– 1	MPa.m$^{1/2}$

Thermal properties

Glass temperature	77	– 102	°C
Maximum service temperature	120	– 140	°C
Thermal conductor or insulator?	Good insulator		
Thermal conductivity	0.16	– 0.2	W/m.K
Specific heat capacity	1660	– 1710	J/kg.K
Thermal expansion coefficient	*37	– 49	μstrain/°C

Electrical properties

Electrical conductor or insulator?	Poor insulator		
Electrical resistivity	*2.1 × 10^{14}	– 7 × 10^{14}	μohm.cm
Dielectric constant	*5	– 6	
Dissipation factor	*0.1	– 0.15	
Dielectric strength	*0.4	– 0.6 10^6	V/m

Wood remains one of the world's major structural materials as well finding application in more delicate objects such as furniture.

Eco properties: material

Embodied energy, primary production	7.4	– 8.2	MJ/kg
CO_2 footprint, primary production	0.45	– 0.49	kg/kg
Water usage	*730	– 1.5e3	l/kg
Eco-indicator	6.3	– 6.9	millipoints/kg

Eco properties: processing

Construction energy	*0.909	– 1.1	MJ/kg
Construction CO_2	*0.045	– 0.05	kg/kg

Recycling

Recycle fraction in current supply	8	– 10	%

Typical uses. Flooring; stairways, furniture; handles; veneer; sculpture, woodenware; sashes; doors; general millwork; framing—but these are just a few. Almost every load-bearing and decorative object has, at one time or another, been made from wood.

Appendix – Useful numbers and conversions

A.1 Introduction

Quantitative analysis needs numbers. Many of those needed to understand and quantify eco-aspects of material production and use are presented in the Chapters of the text.

- Table 2.1: approximate efficiency factors for energy conversion and the associated CO_2 emission per useful MJ.

- Table 6.5: the energy content of fossil fuels and the CO_2 they emit when burnt.

- Table 6.6: the energy efficiency of electricity generation and the related CO_2 per useful kW.hr.

- Table 6.7: the energy and CO_2 costs of alternative modes of transport.

- Table 9.6: the energy and CO_2 rating of cars as a function of mass.

- Chapter 12: data sheets listing the attributes of 47 of the most widely used materials.

This appendix assembles further useful numbers and conversion factors.

A.2 Physical constants in SI units

Physical constant	Value in SI units	
Absolute zero temperature	-273.2	°C
Acceleration due to gravity, g	9.807	m/s^2
Avogadro's number, N_A	6.022×10^{23}	–
Base of natural logarithms, e	2.718	–
Boltzmann's constant, k	1.381×10^{-23}	J/K
Faraday's constant k	9.648×10^4	C/mol
Gas constant, $\bar{R}$	8.314	J/mol/K

Physical constant	Value in SI units	
Permeability of vacuum, μ_0	1.257×10^{-6}	H/m
Permittivity of vacuum, ε_0	8.854×10^{-12}	F/m
Planck's constant, h	6.626×10^{-34}	J/s
Velocity of light in vacuum, c	2.998×10^8	m/s
Volume of perfect gas at STP	22.41×10^{-3}	m^3/mol

A.3 Conversion of units, general

Quantity	Imperial unit	SI unit
Angle, θ	1 rad	57.30°
Density, ρ	1 lb/ft^3	16.03 kg/m^3
Diffusion coefficient, D	1 cm^3/s	1.0×10^{-4} m^2/s
Energy, U	See Section A5	
Force, F	1 kgf 1 lbf 1 dyne	9.807 N 4.448 N 1.0×10^{-5} N
Length, l	1 ft 1 inch 1 Å	304.8 mm 25.40 mm 0.1 nm
Mass, M	1 tonne 1 short ton 1 long ton 1 lb mass	1000 kg 908 kg 1107 kg 0.454 kg
Power, P	See Section A5	
Stress, σ	See Section A4	
Specific heat, Cp	1 cal/gal.°C Btu/lb.°F	4.188 kJ/kg.°C 4.187 kg/kg.°C
Stress intensity, K_{1c}	1 ksi $\sqrt{in}$	1.10 MN/m$^{3/2}$
Surface energy γ	1 erg/cm^2	1 mJ/m^2
Temperature, T	1°F	0.556°K
Thermal conductivity λ	1 cal/s.cm.°C 1 Btu/h.ft.°F	418.8 W/m.°C 1.731 W/m.°C
Volume, V	1 Imperial gall 1 US gall	4.546×10^{-3} m^3 3.785×10^{-3} m^3
Viscosity, η	1 poise 1 lb ft.s	0.1 N.s/m^2 0.1517 N.s/m^2

A.4 Stress and pressure

The SI unit of stress and pressure is the N/m^2 or the Pascal (Pa), but from a materials point of view it is very small. The levels of stress large enough to distort or deform materials are measured in megaPascals (MPa). The table list the conversion factors relating MPa to measures of stress used in the older cgs and metric systems ($dyne/cm^2$, kgf/mm^2) by the Imperial system (lb/in^2, ton/in^2) and by atmospheric science (bar).

Conversion of units – stress and pressure

To → From ↓	MPa	dyn/cm²	lb/in²	kgf/mm²	bar	long ton/in²
			Multiply by			
MPa	1	10^7	1.45×10^2	0.102	10	6.48×10^{-2}
dyn/cm²	10^{-7}	1	1.45×10^{-5}	1.02×10^{-8}	10^{-6}	6.48×10^{-9}
lb/in²	6.89×10^{-3}	6.89×10^4	1	7.03×10^{-4}	6.89×10^{-2}	4.46×10^{-4}
kgf/mm²	9.81	9.81×10^7	1.42×10^3	1	98.1	63.5×10^{-2}
bar	0.10	10^6	14.48	1.02×10^{-2}	1	6.48×10^{-3}
long ton/ in²	15.44	1.54×10^8	2.24×10^3	1.54	1.54×10^2	1

A.5 Energy and power

The SI units of energy is the Joule (J), that of energy the Watt (W = 1 J/sec), or multiples of them like MJ or kW. If energy and power were always listed in these units, life would be simple, but they are not. First there are Imperial units Btu, ft.lbf, and ft.lbf/s. Then there are units of convenience: kWhr for electric power, hp for mechanical power. There are the units of the oil industry: barrels (7.33 barrels = 1 tonne, 1000 kg) and toe (tonnes of oil equivalent – the weight of oil with the same energy content). Switching between these units is simply a case of multiplying by the conversion factors listed below.

Conversion of units – energy*

To →	MJ	kWhr	kcal	Btu	ft lbf	toe
From ↓				**Multiply by**		
MJ	1	0.278	239	948	0.738×10^6	23.8×10^{-6}
kWhr	3.6	1	860	3.41×10^3	2.66×10^6	85.7×10^{-6}
kcal	4.18×10^{-3}	1.16×10^{-3}	1	3.97	3.09×10^3	99.5×10^{-9}
Btu	1.06×10^{-3}	0.293×10^{-3}	0.252	1	0.778×10^3	25.2×10^{-9}
ft lbf	1.36×10^{-6}	0.378×10^{-6}	0.324×10^{-3}	1.29×10^{-3}	1	32.4×10^{-12}
toe	41.9×10^3	11.6×10^3	10×10^6	39.7×10^6	30.8×10^9	1

MJ = megajoules; kWhr = kilowatt hour; kcal = kilocalorie; Btu = British thermal unit; ft lbf = foot-pound force; toe = tonnes oil equivalent.

Conversion of units – power*

To →	kW (kJ/s)	kcal/s	hp	ft lbf/s
From ↓		**Multiply by**		
kW (kJ/s)	1	4.18	1.34	735
kcal/s	0.239	1	0.321	176
hp	0.746	3.12	1	545
ft lbf/s	1.36×10^{-3}	5.68×10^{-3}	1.82×10^{-3}	1

kW = kilowatt; kcal/s = kilocalories per second; hp = horse power; ft lb/s = foot-pounds/second

A.6 Fuels

The energy content and CO_2 emission of fossil fuels are listed in Table 6.5 of Chapter 6. Here we list alternative measures of quantity for oil and gas. Coal is always quantified in tonnes or short tons.

Measures of quantity

Crude oil	Natural gas
1 barrel = 35 Imperial gallons	1 billion m^3 (10^9 m^3) = 35.5×10^9 ft^3
= 42 US gallons	= 6.29×10^6 boe*
= 159 liters	= 0.9×10^6 toe**
≈135 kg	= 0.73×10^6 tonnes LNG***

*boe = barrel of oil equivalent; **toe = tonne of oil equivalent; ***LNG = liquid natural gas.*

A.7 Energy prices (2007 data)

Energy prices fluctuate. All, ultimately, are tied to the price of oil, which in 2007 was $60 per barrel.

Approximate energy prices, 2007 data

Energy form	Price, usual units	Price per MJ
Electricity, industrial	$ 0.06/kWhr	$ 0.017/MJ
Electricity, grid	$ 0.11/kWhr	$ 0.031/MJ
Heavy fuel oil	$ 335/tonne	$ 0.0073/MJ
Industrial gas	$ 282/10^7 kilocals	$ 0.0067/MJ
Coal	$ 40/tonne	$ 0.0025

A.8 Further reading

Carbon Trust (2007), "Carbon footprint in the supply chain", (www.carbontrust.co.uk)

Congressional Budget Office, US Congress (1982), "Energy use in freight transportation", Washington DC. *(Energy per ton-mile for transport.)*

Electricity Information (2008), IEA publications, ISBN 978-9264-04252-0. *(One of a series of IEA statistical publications, this one giving every statistic you ever wanted to know and plenty you don't about electricity.)*

http://www.simetric.co.uk/sibtu.htm *(SI to Imperial conversion tables.)*

International Chamber of Shipping (2005), International Shipping Federation, Annual review, www.marisec.org/annualreview/annualreview.pdf. *(Energy per ton-mile for shipping.)*

Jancovici, J-M. (2007), http://www.manicore.com *(A mine of useful information about energy.)*

MacKay, D.J.C. (2008), "Sustainable energy – without the hot air", UIT Press, Cambridge, UK. ISBN 978-0-9544529-3-3. *(Helpful assemble of useful and quirky numbers relevant to energy generation and use.)*

Network Rail (2007), www.networkrail.co.uk/freight/ *(Energy per tone.km for rail transport.)*

Nielsen, R. (2005), "The little green handbook", Scribe Publications Pty Ltd, Carlton North, Victoria, Australia. ISBN 1-9207-6930-7 *(Well researched tables of energy information.)*

Shell Petroleum (2007), "How the energy industry works", Silverstone Communications Ltd, Towchester, UK. ISBN978-0-9555409-0-5. *(Both BP and Shell publish annual compilations of energy statistics.)*

Transport Watch UK (2007), www.Transwatch.co.uk/Transport-fact-sheet-5 *(More on energy for transport.)*

USGS (2007), Minerals Information: Mineral commodity summaries. http://minerals.usgs.gov/minerals/pubs/commodity/ *(The bible of resource data.)*

Index

The materials teaching resource

CES EduPack™ from Granta Design is the world-leading teaching resource for materials in engineering, science, processing, and design. It complements this book and is used at hundreds of universities and colleges worldwide.

CES EduPack offers: extensive materials information, including eco property data; software, including eco audit and materials selection capabilities *(pictured)*; lectures, including on eco design topics; and student projects and exercises.

With CES EduPack you can:

- **Get more from this text.** Students can complete supporting exercises as they read and explore the eco impact of materials in depth. Together, the book and software offer a superb integrated learning resource.

- **Meet your learning and teaching needs.** CES EduPack is very flexible. Different databases and supporting resources match varying needs — from first to final year and across engineering disciplines. The Eco Design Edition provides everything that you need to complement this book.

CES EduPack is normally purchased by a college or university and then made available to students. Teachers - contact Granta Design. We'll be happy to arrange an online demonstration of the software and to tell you about the supporting lectures, projects, and exercises.

GRANTA
MATERIAL INSPIRATION

www.grantadesign.com | info@grantadesign.com

Eco design in industry?

Are you trying to minimize the environmental impact of your company's products and new designs?

Granta can help, with a range of software tools to manage critical materials information and to apply it in making design decisions. Minimize energy usage, CO_2 footprint, and other wastes and emissions. Avoid restricted substances. Plan for product end-of-life.

Find out more and read about the work of Granta's Environmental Materials Information Technology (EMIT) Consortium:

grantadesign.com/eco/